H. J. Campbell

Textbook of Elementary Biology

H. J. Campbell

Textbook of Elementary Biology

ISBN/EAN: 9783337216528

Printed in Europe, USA, Canada, Australia, Japan

Cover: Foto ©berggeist007 / pixelio.de

More available books at **www.hansebooks.com**

TEXT-BOOK

OF

ELEMENTARY BIOLOGY

BY

H. J. CAMPBELL, M.D. LOND.

*Senior Demonstrator of Biology and Demonstrator of Physiology in the
Medical School of Guy's Hospital*

WITH A HUNDRED AND THIRTY-SIX ILLUSTRATIONS

LONDON:

SWAN SONNENSCHEIN & CO.

NEW YORK: MACMILLAN & CO.

1893.

PREFACE.

———◆———

IN the following pages, I have endeavoured to give a short account of some of the more important facts of Biology. I have dealt at some length with the subjects of Protoplasm, Cells, Cell-division, Reproduction, the early stages of Development, and the massing together of cells to form Tissues, as these subjects are of fundamental importance not only for the study of Zoology and Botany, but also for that of Physiology and Pathology. The general chapters upon Vertebrata, Invertebrata, and Plant Structure are very brief, and the student is recommended, should he wish to gain a fuller knowledge of these subjects, to refer to the more detailed accounts of them given in the larger text-books. Among the types which are somewhat fully described in the latter part of the book, I have included all those selected by the Conjoint Board of the Royal College of Physicians and Surgeons for their examination in Elementary Biology. To these I have added a short chapter on the Dog-fish, as a type of the lower Craniata. In reading the book for the first time, the student is advised, both to study the chapters in the order laid down on page xii., and also to read only the paragraphs in large print, those in small type being intended only as amplifications. In order to obtain clear ideas of Biology, it is absolutely necessary that reading should be accompanied by the dissection and examination of specimens. The student cannot be too strongly urged to devote as much time as possible to practical work.

I have made frequent reference to the following books : Thompson's *Outlines of Zoology*, Marshall and Hurst's *Practical Zoology*, Claus and Sedgwick's *Text-book of Zoology*, Leuckart's *Human Parasites*, Prantl and Vines' *Text-book of Botany*, and Parker's *Elementary Biology*. In addition to these I have referred to numerous papers by Professor Ray Lankester, Mr. W. E. Hoyle, Mr. F. E. Beddard, and many others. To all these authors I take this opportunity of expressing my grateful acknowledgments. The sources from which the diagrams have been taken are in all cases mentioned. I wish especially to thank Messrs. Smith, Elder, & Co., and Mr. Pentland, for their kind permission to insert diagrams from books published by them. I have further to thank Mr. F. E. Beddard for numerous suggestions, and also for references to original papers, and Dr. Washbourne for his friendly criticism of the chapter on Bacteria.

CONTENTS.

CHAP.   PAGE

    PREFACE v

    LIST OF ILLUSTRATIONS ix

  I. INTRODUCTION 1

  II. PROTOPLASM 3

  III. CELLS 10

  IV. CELL DIVISION 20

  V. REPRODUCTION AND THE EARLY STAGES OF DEVELOP-

    MENT 23

  VI. TISSUES . . 40

  VII. ANIMAL TISSUES . 43

  VIII. PLANT TISSUES 52

  IX. GENERAL REVIEW OF PLANT STRUCTURE. . 63

  X. DIFFERENCES BETWEEN PLANTS AND ANIMALS 79

  XI. GENERAL REVIEW OF THE INVERTEBRATA 83

  XII. GENERAL REVIEW OF THE VERTEBRATA . 120

  XIII. AMŒBA 156

  XIV. YEAST PLANT (TORULA OR SACCHAROMYCES CEREVISIÆ) 161

  XV. PROTOCOCCUS VIRIDIS; PROTOCOCCUS OR HÆMATO-

    COCCUS PLUVIALIS; GLEOCAPSA 164

  XVI. BACTERIA 170

viii · CONTENTS.

CHAP.		PAGE
XVII.	VORTICELLA	176
XVIII.	GREGARINÆ	183
XIX.	HYDRA	188
XX.	THE LIVER-FLUKE (FASCIOLA HEPATICA)	197
XXI.	TAPE-WORMS (CESTODES)	208
XXII.	ROUND WORMS (NEMATODES)	229
XXIII.	THE MEDICINAL LEECH (HIRUDO MEDICINALIS)	241
XXIV.	THE DOG-FISH (SCYLLIUM CANICULA)	252
XXV.	GENERAL REVIEW OF THE MAMMALIA	256
	INDEX AND GLOSSARY	267

LIST OF ILLUSTRATIONS.

FIG.		PAGE
1.	A cell from a staminal hair of Tradescantia virginica	7
2.	Immature egg-cell of an Echinoderm as a typical animal cell	10
3.	Various forms of plant cells	11
4.	Plant cells of different stages of growth	14
5.	Spirogyra showing chromatophores, pyrenoids, etc.	16
6.	Chlorophyll corpuscles	18
7.	Cell division	21
8.	Egg-cell	24
9.	Formation of polar vesicles in Asterias glacialis	25
10.	Spermatozoa of animals	26
11.	Young ova and sperm mother-cells	27
12.	Egg-cell in process of fertilisation showing male and female pronuclei	27
13.	Fertilised egg-cell	27
14.	Unequal segmentation of Frog's egg	28
15.	Diagram of an egg with protoplasm forming a germ disc	29
16.	Segmentation of the germinal disc of a Hen's egg	30
17.	Blastula of Amphioxus	31
18.	Blastula of Triton tæniatus	31
19.	Formation of the gastrula in Amphioxus	32
20.	Diagram of Hen's egg	32
21.	Section through the germ disc of a Hen's egg	33
22.	Longitudinal section through the germ disc of a Finch's egg	33
23.	Surface view of the area pellucida in the blastoderm of a Chick	34
24.	Optical sections of a segmenting Rabbit's egg in two stages	35
25.	Rabbit's egg 70 to 90 hours after fertilisation	36
26.	Diagram to show the derivation of the Cœlom	38
27.	Diagram of the development of the alimentary canal of a Chick	39
28.	Epithelial cells	44
29.	White fibrous tissue	45
30.	Hyaline cartilage	46
31.	Transverse section of bone	46
2.	Striated muscle	48

FIG.		PAGE
33.	Non-striated muscle	49
34.	Nerve cells	50
35.	Hairs on young ovary of Cucurbita	53
36.	Transverse section of leaf showing a Stoma	54
37.	Transverse section of petiole of Helleborus	55
38.	Transverse and longitudinal sections of fibrovascular bundle of Sunflower	57
39.	Transverse section of a root of Acorus calamus	59
40.	Longitudinal section of root apex of Hordeum vulgare	61
41.	Sexual and asexual reproduction in Eurotium repens	64
42.	Thallus of Fucus vesiculosus	65
43.	Nitella flexilis and Chara fragilis	67
44.	Development of the sporogonium of Funaria hygrometrica	68
45.	Sporangia of Scolopendrium vulgare	69
46.	Prothallium of a Fern	70
47.	Antheridium of Adiantum capillus veneris	71
48.	Archegonium of Polypodium vulgare	72
49.	Prothallium and young asexual plant of Adiantum	72
50.	Diagram of a flower	73
51.	Different forms of pistils in flowers	74
52.	Diagrams of hypogynous, perigynous and epigynous flowers	75
53.	Ovuliferous scale of Pinus sylvestris	76
54.	Longitudinal section through ovule of Picea vulgaris	77
55.	Growing cells of Yeast	80
56.	Longitudinal section through a Sponge (Sycon raphanus)	86
57.	Branch of an Obelia stock	88
58.	Branch of a Polyparium of Corallium rubrum	89
59.	Distomum hæmatobium (Bilharzia hæmatobia)	92
60.	Lumbricus rubellus	93
61.	Generative organs of Lumbricus	95
62.	A Star-fish (Echinaster sentus)	97
63.	Diagram of the water vascular system of a Star-fish	98
64.	Development of a Feather-star (Antedon rosacea)	101
65.	Nauplius and adult stages of Sacculina purpurea and pupa of Lernæodiscus porcellanæ	103
66.	Longitudinal section through a Cray-fish	104
67.	Gryllus campestris	106
68.	Digestive apparatus and renal tubes of the Honey bee	107
69.	Larva of Ephemera showing tracheal gills	109
70.	Formica herculanea and Formica rufa	111
71.	Female spiders	113
72.	Demodex folliculorum	114
73.	Pentastomum denticulatum	115
74.	Swan mussel (Anodon)	117
75.	Head and anterior region of skeleton of Dog-fish	124

FIG.		PAGE
76.	Skeleton of an Egyptian vulture	126
77.	Skeleton of hand of Orang, Dog, Pig, Ox, Tapir, and Horse	128
78.	Median longitudinal section of a Sheep's skull	129
79.	Alimentary system of a Bird	132
80.	Alimentary canal of Man	134
81.	Longitudinal section of anterior half of a Cat	136
82.	Blood corpuscles of various Animals	138
83.	Diagram of the circulatory organs of a Fish	139
84.	Aortic arches of the Tadpole	141
85.	Diagram of the circulatory system in a higher Vertebrate	143
86.	Diagram of the renal excretory organs (nephridia) of a Worm	145
87.	Diagram of the renal excretory organs of a Dog-fish embryo	146
88.	Ciliated funnel and Malpighian body from kidney of Proteus	147
89.	Diagrams of the brain of a Human embryo	148
90.	Brain and anterior part of spinal cord of a Fish (Hexanchus griseus)	149
91.	Nervous system of the Frog	151
92.	Female generative organs in Mammals	153
93.	Urinary and sexual organs in a Mouse	154
94.	Amœbæ	157
95.	Growing cells of the Yeast plant	161
96.	Protococcus viridis	165
97.	Gleocapsa	168
98.	Bacteria of the fur of teeth	170
99.	Bacteria, isolated and in zooglœa condition	173
100.	Vorticella	178
101.	Vorticella microstoma showing fission and conjugation	181
102.	Gregarina	185
103.	Diagrammatic longitudinal section of Hydra	189
104.	Longitudinal section through body-wall of Hydra	191
105.	Liver-fluke from the ventral surface	199
106.	Liver-fluke from the ventral surface showing reproductive organs	202
107.	Five stages in the life history of the Liver-fluke	204
108.	Head of Tænia solium	209
109.	Half-ripe and ripe joints of Tænia solium	209
110.	Two joints of Tænia solium showing uterus	209
111.	The development of Tænia solium to the Cysticercus stage	211
112.	Measly pork	212
113.	Bothriocephalus latus	216
114.	Echinococcus, developmental stages	217
115.	Tænia mediocanellata (saginata)	219
116.	Transverse section of a joint of Tænia mediocanellata	221
117.	Longitudinal section of a joint of Tænia mediocanellata	222
118.	Head of Tænia serrata showing excretory vessels	223
119.	Proglottis of Tænia mediocanellata showing reproductive organs	225

living matter grows by intussusception, that is to say, the new material is taken up by and becomes inseparably combined with the old.

3. Its capacity for reproducing itself.—In living matter the race is perpetuated by the detachment of portions, which tend to run through the same cycle of forms as the parent.

In **Biology**, chiefly for the sake of convenience, a broad distinction is made between plants and animals, the study of plants being termed **Botany**, and that of animals **Zoology**. However, as it will be seen later, it is by no means easy to define the line of distinction between plants and animals, especially in their lowest forms.

CHAPTER II.

PROTOPLASM.

P ROTOPLASM (πρῶτος, first; πλάσμα, anything formed), or as it has been more appropriately called Bioplasm, is a viscid, semi-fluid substance, exceedingly complex in its chemical composition. Its constitution cannot be represented by any chemical formula, as no chemical analysis can be made of it. It cannot, of necessity, be analysed while alive, as it is in such a state of unstable equilibrium, that the very attempt to analyse it at once causes its disintegration, and death so materially alters its composition and the arrangement of its molecules, that it practically is no longer protoplasm.

Protoplasm then is a living whole whose molecules, being in a most unstable condition, are very liable to assume other relative positions. What is then the force which binds these molecules together, as it were in spite of themselves ? We do not know. For want of a better name we call it Vital Force. When it is overcome, death ensues, and we deal no longer with living, but with dead matter, which as such may be subjected to analysis and may have its composition determined. We must not forget, however, that then its whole character is altered ; we are no longer dealing with a single living substance, but with a mixture of dead substances.

One of the most striking characteristics of protoplasm is its irritability or excitability. By this is meant the readiness with which it reacts to stimuli.   A force, which is not powerful enough to kill it, and thus to cause its complete disintegration, will yet cause some change in the relative positions of its molecules.   For instance, a strong current of electricity will kill protoplasm, whereas a weak one will only cause it to contract.

Another very striking characteristic of protoplasm is its need of oxygen.   It requires a constant supply of this gas to maintain its activity.   This great greed for oxygen of necessity makes protoplasm a strong reducing or deoxidising agent.

The oxygen which protoplasm is continually absorbing, is as continually being given out again combined with carbon in the form of carbon dioxide.

Thus we have observed four of the most important characteristics of protoplasm.

1. **Its instability.**
2. **Its irritability.**
3. **Its reducing or deoxidising power.**
4. **Its power of giving up carbon dioxide.**

CHEMICAL NATURE OF PROTOPLASM.

Dead protoplasm can be analysed, and its composition determined.   It is found to consist of proteid, water, carbohydrates, fat, and salts.   Of these by far the most abundant, and the most important, are the proteid and the water.   The greater the activity of the protoplasm during life, the greater will be the amount of water which can be extracted from it after death.

The proteid contained in dead protoplasm is a complex substance, the composition of which we are unable to express in the terms of a chemical formula, as its constitution is very varied.   In fact there appears to be a group of substances all classed together under the common name of proteid, which exhibit very different properties.   They all consist of carbon, hydrogen, nitrogen, oxygen, and sulphur; but these elements are not always combined together in the same proportions.   The variations, however, are within limits; the amount of carbon varies from $51 \cdot 5\%$ to $54 \cdot 5\%$, that of hydrogen from 6·9% to 7·3%, that of nitrogen from 15·2% to 17%, that of oxygen from 20·9% to 23·5%, and that of sulphur from 0·3% to 2%.

Carbohydrates are substances consisting of carbon, hydrogen, and oxygen, the two latter always being present in the proportion in which they occur in water (H_2O). Thus a molecule of starch consists of 6 atoms of carbon, 10 of hydrogen, and 5 of oxygen ($C_6 H_{10} O_5$).

Fats also consist of carbon, hydrogen, and oxygen; but there is far more hydrogen in proportion to the oxygen than in water.

The salts vary considerably, both in quantity and in nature, the following elements being very frequently found: Potassium, sodium, magnesium, iron, and phosphorus.

It will be observed that out of the various constituents of dead protoplasm, the only one which contains nitrogen is the proteid.

During the whole of its life, protoplasm is, on the one hand, constantly assimilating food, and building itself up; and, on the other hand, as constantly breaking down, and getting rid of its waste products.

Both the foods and the waste products are of less complex structure than the protoplasm itself. One of the most important food substances is oxygen, and one of the commonest waste products is carbon dioxide. This elimination of carbon dioxide, as has already been observed, is one of the most marked characteristics of protoplasm.

The life history of protoplasm may be represented by a curved line. At the one end are the food substances, on their way to form protoplasm; at the summit is the active living protoplasm; at the other end are the waste products, and the dead protoplasm.

The chemical changes which take place during the life history of protoplasm are collectively spoken of as its metabolism, that is, the changing over from one state to another. The building up, synthesis, or assimilation of non-living matter is described as its anabolism, and the breaking down of the living matter again into dead matter as its katabolism.

Professor Gaskell considers that the constructive or anabolic process of restitution goes on constantly and of itself—*i.e.*, without the necessity of stimulus ; whereas the katabolic or disruptive process, in which energy is discharged, takes place only occasionally, and in obedience to stimulus. This theory suggests an alternation of rest with activity, of self-regulative construction with stimulated disruption.

The importance of this liberation of energy during the disruptive process cannot be over estimated, for evidence is continually accumulating to show that to this process can be ascribed all the so-called vital phenomena exhibited by the protoplasm, the changes in form, the contractions, the streaming movements, etc. ; and this also renders it easy to understand the terms excitability or irritability, which have been before made use of. The unstable protoplasm is readily affected by external stimuli, that is, the disruptive or katabolic process is readily induced, and the changes in form, etc., in the protoplasm are but the external manifestations of the processes which are taking place within the protoplasm itself.

If one of the hairs from the base of a stamen of a flower of the Tradescantia virginica (Spiderwort) be examined with a high power of the microscope, it will be seen to be composed of a number of cells. If one of these cells be carefully examined, the greater part of it will be seen to consist of a violet-coloured transparent fluid, the cell-sap; but surrounding the cell-sap, and hence lining the inside of the cell-wall, there is a layer of semi-fluid substance containing numerous granules. This granular semi-fluid substance is protoplasm. At one part, imbedded in it, a denser body will be seen; this is the nucleus. Occasionally nucleoli may be distinguished in the nucleus itself. Threads of protoplasm radiate from the nucleus and stretch to the lining of the cell-wall.

If the protoplasm be closely watched, it will be seen that streaming movements are constantly taking place in it, which are accompanied by a slow but constant change in the shape of the protoplasm itself. Although these movements can only be observed in a few cells, it is probable that they occur in all at one stage or other of their development.

Most animal cells are so closely packed together that they cannot move visibly until they are set free. If however a cell of the embryo of the fowl or of the frog be set free at a certain stage of its existence, it may be observed to change its shape slowly, now protruding one part of its substance and now another.

These movements may, however, be more easily observed in the colourless blood corpuscles of the Frog or of Man. They are described as amœboid movements, from their similarity to those exhibited by the Amœba, one of the lowest forms of animal life.

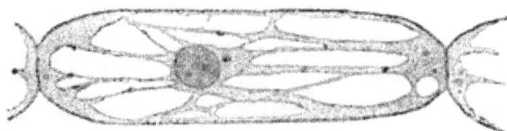

Fig. 1.—A cell from a staminal hair of Tradescantia virginica (from Strasburger).

Although so characteristic of protoplasm, they can, under certain conditions, be made to occur in non-living matter. Professor Bütschli has succeeded in inducing movements in droplets of thick olive oil suspended in a weak solution of potassium carbonate, which closely resemble the movements of the protoplasm in the Amœba, and under suitable conditions these movements may be seen to occur continuously for a considerable period of time.

As these movements of protoplasm form one of the most delicate and charcteristic signs of its life, so their cessation forms one of the most reliable, and most easily recognisable evidences of its death.

It has already been stated that protoplasm is constantly absorbing oxygen, and as constantly eliminating carbon dioxide. The two processes are described as the **respiration of protoplasm,** and as protoplasm occurs in every living cell of both plants and animals, it of necessity follows that the processes of respiration must go on in every living organism. The oxygen may be withheld for a short time without killing the protoplasm, which only becomes dormant, and recovers itself when the oxygen is again supplied, but, if active protoplasm be kept too long without oxygen, it invariably dies.

> This is only the case in protoplasm in which active processes are taking place. The protoplasm which is lying quiescent, such as that in seeds, does not die, but it requires oxygen as soon as growth commences.

It is also necessary that the cell should be able to eliminate its carbon dioxide.

Two other conditions are necessary if the living processes of protoplasm are to be actively carried on ; namely, that it should be able constantly to obtain a sufficiency of **water**, and that the **temperature** should not fall below 4° C., nor rise above 40° C. Changes of temperature within these limits affect the rapidity of the protoplasmic movements, a rise of temperature quickening them, while a fall slackens them.

Electricity also affects the movements, a weak current causing them to increase in rapidity, while a strong one kills the protoplasm. Similarly, chemical reagents if very dilute, and mechanical stimuli if weak, also excite these movements, whilst strong reagents and stimuli cause death in the protoplasm.

All the movements which a body makes or exhibits are due to the action of these stimuli on the protoplasm in the individual cells which form the organism.

HISTORICAL.

> Rösel von Rosendorf, in 1755, gave a description of the Amœba, which he called the Proteus Animalcule. He dwelt at some length upon the form and movements of what we now call living protoplasm. Later on, Robert Brown, Schwann and Schleiden (1831 — 1839) described the structure, origin, and functions of the cell substance. Unfortunately, however, they retarded rather than advanced our knowledge, for all their descriptions were coloured by the theory, since disproved, that the cell-wall is the most important part of the cell.

> Dujardin in 1835 described the substance which we now call protoplasm

under the name of sarcode, and it was the careful work of this observer that first directed enquiry along the lines which it has since followed.

Hugo von Mohl in 1846, in his description of the vegetable cell, first clearly distinguished protoplasm not only from cell-wall but also from cell-sap, and it was he who gave to it the name by which it is still known. Max Schulze in 1861 identified the vegetable protoplasm of Mohl with the animal sarcode of Dujardin.

CHAPTER III.

CELLS.

L EYDIG defined a cell as "a mass of **protoplasm** furnished with a **nucleus**." This definition holds true in the vast majority of cases, but it is not universally true, for there are a certain number of minute unicellular plants and animals, each of which apparently consists of nothing but a single droplet of protoplasm, in which, so far, no nucleus has been

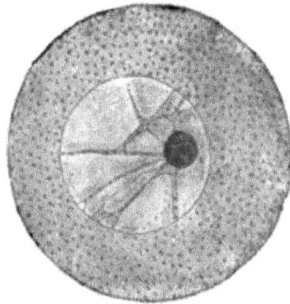

FIG. 2.—Immature egg-cell from the ovary of an Echinoderm, showing nucleus, nucleolus, and nuclear network (from Hertwig).

discovered. Almost all plant cells have, moreover, in addition, a **cell-wall**, which surrounds and encloses the protoplasm.

Many animal cells are also enclosed by a **cell-membrane** ; so that the older definition of a cell " that it consists of a mass of protoplasm surrounded by a cell-membrane, and that it contains a nucleus," is true in a large proportion of cases.

The cell-membrane or wall is formed by substances which, in many cases, are waste products of the protoplasm, deposited either in the external layer of the protoplasm or upon the outside of it.

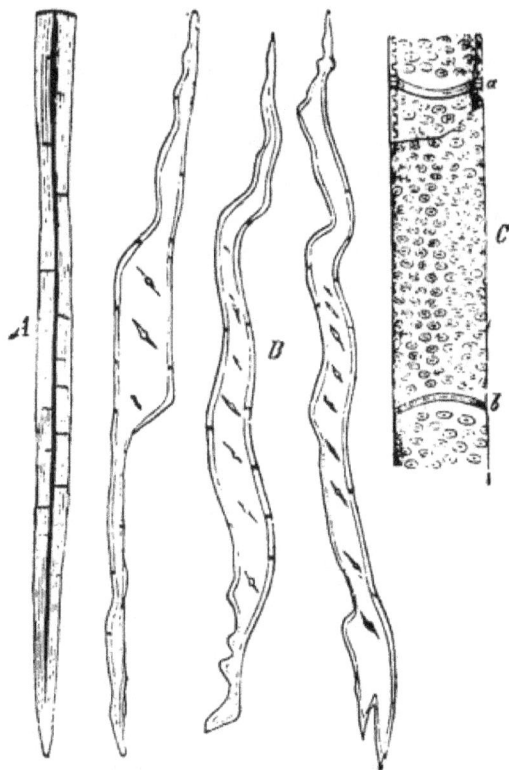

FIG. 3.—Various forms of plant cells (from Prantl). *A*, the end of a bast fibre with strongly thickened pitted walls (longitudinal section) ; *B*, wood cells, surface view and section; *C*, part of vessel with bordered pits, cut open at the top. At *a* and *b* the remains of the absorbed septa are visible.

The number of cells, in which nuclei have not been discovered, is steadily diminishing ; in some cases of cells without nuclei, the nuclear substance seems to be diffused through the protoplasm. There is no doubt that the nucleus is a very important part of the cell, but exactly what are its functions and relative importance is not yet fully understood. There is no doubt that the nucleus plays an important part in cell division, as also in the processes of fertilisation. If a cell loses its nucleus it soon dies. The nucleus frequently contains within itself one or more minute rounded bodies called nucleoli, and a description has been given of a similar smaller body within the nucleolus to which the name of endonucleolus has been given.

Cells vary very considerably in size, some being less than $\frac{1}{3000}$ inch in diameter, whilst others can be seen with the naked eye ; they also differ very much from one another in shape, as may be seen from the figures of animal cells in figs. 28 and 30.

So also very wide differences are met with in the shape of the cells in differents parts of a plant, as seen in the figures on pp. 7, 11, and 14.

Every cell has a life history of its own. It increases in size until it reaches its adult condition ; later on its activity diminishes, and at last it dies or divides. All cells resemble each other when they are first formed ; it is only as they increase in size and begin to approach maturity that they take on the special forms which fit them to carry out their special functions in the economy of the plant or animal. As the functions performed by the cells differ widely from each other, it is only to be expected that the mature cells should also differ much from each other, and from their own youngest state.

The protoplasm of the cell until recently was generally considered to be completely homogeneous, and although it might sometimes contain watery fluid (vacuoles) imbedded in it, yet it was supposed to be entirely devoid of structure. This may possibly be true in some cases ; but it is more generally true, especially in the case of fixed animal cells, that the protoplasm consists of two parts—a more fluid part, which is clear and homogeneous, and has been called the hyaloplasm ; and a denser, more solid part, which forms a network of fibrils, the reticulum or spongioplasm. As a rule the older the cell, the greater is the relative amount of reticulum that it contains, and conversely the younger the cell, the more of hyaloplasm is there in it. The massing together of reticulum at the points where the fibrils meet, is supposed by some authorities to give the appearance of minute granules, which is so constantly seen in protoplasm.

In cells where no cell-membrane is present, the peripheral layer of protoplasm is usually clearer than that in the interior, the difference being sufficiently well marked to justify the use of the terms **ectoplasm** and **endoplasm**, the former being applied to the clearer external layer, the latter to the more granular internal portion. It is in the endoplasm that foreign substances occur, such as granules of proteid or starch, globules of fat, crystals of various kinds, pigment granules and globules of watery fluid (vacuoles).

The reticulum forms a sort of framework to support the hyaloplasm. The active streaming movements of the protoplasm take place in the hyaloplasm.

Some cells, both animal and vegetable, in which no reticulum has been seen, are supposed to consist of hyaloplasm alone ; if that be the case then it must be the hyaloplasm which is the essential part of the cell-protoplasm.

As the form and appearance of the adult animal cell correspond in the main with the above description of the typical cell, whereas that of the adult plant cell differs considerably, I have thought it well to give a somewhat detailed account of the latter, including the most important facts concerning the structure of the cell-wall, the constitution of cell-sap, and the special substances contained in the protoplasm of plant cells.

In the majority of cases, plant cells, especially those of the higher plants, are distinguished from animal cells by their possessing, a **firm, elastic cell-wall**, composed of **cellulose**, a substance allied to starch in its chemical structure. In such cells, a **layer of protoplasm** lies in contact with the inner surface of the wall ; and the **nucleus**, a more solid rounded body, is imbedded in this layer. The space in the centre of the cell, **the vacuole**, is filled with a watery fluid, the **cell-sap**. Young cells

contain no vacuoles, the cell-sap being diffused through the
protoplasm which fills the cell. As the cell develops, the

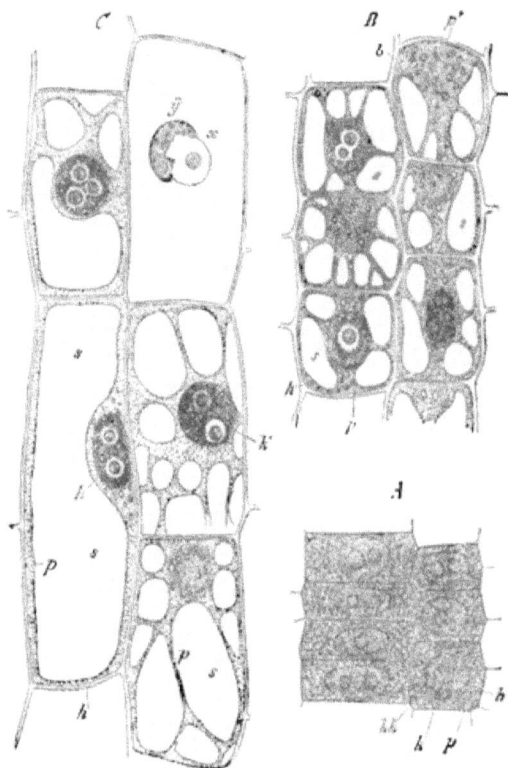

Fɪɢ. 4.—Parenchyma cells from the cortical layer of the root of Fritillaria imperialis ;
longitudinal sections. *A*, very young cells still without cell-sap ; *B*, older
cells. The cell-sap *s* forms separate drops in the protoplasm *p* ; *C*, still older
cells ; *h*, cell-wall ; *k*, nucleus (after Sachs).

cell-sap collects in small drops at different parts of the cell ; as
these drops increase in size, they coalesce, and at the same
time the bands of protoplasm which separate them become
absorbed into the peripheral layer.

It is important to remember that the essential parts of the cell are the protoplasm and the nucleus ; the cell-sap and the cell-wall are both products of the activity of this substance.

As the cell-wall grows its chemical composition frequently changes more or less, the three most important changes being (1) into wood, (2) into cork, and (3) into mucilage. The wood, such as that composing the stems of the forest trees, is easily permeated by water ; the cork, which forms the bark of the trees, is elastic and not easily permeated by water, whilst the mucilage is formed by the cell-wall swelling up very considerably in water and becoming soft and gelatinous. This is well seen in some seaweeds.

Mineral matters sometimes occur in large quantities in the walls of cells. The commonest of these are silica and calcium carbonate. Silica is present in considerable quantities in such rigid structures as stems of grasses and in the leaves of the beech.

Cell-sap.—The cell-sap, which saturates the whole of the cell-wall and protoplasm and which fills the vacuoles, consists of a solution of various substances formed by the activity of the protoplasm, such as sugar, soluble salts, vegetable acids, and colouring matters. The colours of many flowers are due to these colouring matters in the cell-sap. The soluble salts are often crystallised out from the cell-sap in the cavity or the wall of the cell.

Substances contained in the protoplasm.—The most important substances found in protoplasm, and which are formed by the activity of this substance, are chlorophyll, crystalloids, aleurone grains, fat, and starch.

The green colour of most parts of the higher plants is due to the presence of chlorophyll corpuscles in the cells. A chlorophyll corpuscle is a small mass of specialised protoplasm.

It consists of a colourless ground substance, through which is diffused a small quantity of a green colouring matter called chlorophyll, which is soluble in alcohol.

In some of the lower plants the chlorophyll does not occur in special corpuscles, but is diffused through the whole of the protoplasm. In others the protoplasm which contains the chlorophyll occurs in the form of spiral bands.

The chlorophyll corpuscles grow in the manner characteristic of living substances, that is, by the intussusception of new

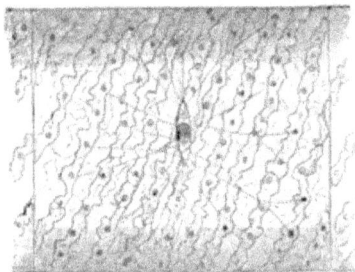

Fig. 5.—Spirogyra (from Strasburger). A cell of a thread showing chlorophyll-bands, pyrenoids, and nucleus with its suspending threads.

matter, the new particles being deposited amongst those already present. They multiply by each corpuscle dividing into two.

For the development of the green colouring matter light is essential ; if a plant be grown in darkness, the corpuscles will be formed, but they will not be green. It is also necessary for the formation of chlorophyll that the plant should be supplied with iron.

In some plants, as in the Bladder wrack, the green colour is obscured by the presence of a differently coloured cell-sap. If this sap be dissolved out by soaking the plant for a considerable time in cold fresh water, the green colour may be brought into view.

Other colouring matters occur in the cell-saps of various coloured plants: the red colouring matter is phycoerythrin, the olive green is phycophæin, and the bluish green phycocyan.

The composition of chlorophyll is unknown, except that it is an iron-containing substance.

Many plants turn brown or yellow in the autumn. This change of colour is caused by most of the chlorophyll granules becoming dissolved and the remaining ones turning yellow.

When fruits ripen, changes take place in the chlorophyll, the green colouring matter frequently becoming first yellow and then red, while the form of the corpuscle becomes altered. The yellow colour of many flowers, as, for instance, that of the Dandelion, is due to the presence of protoplasmic bodies containing a yellow colouring matter closely allied to chlorophyll. The colour of the Virginian creeper at the end of summer is due to the presence of other colouring matters in the cell-sap.

Chlorophyll is of the utmost importance for the nutrition of the plant, for by its help in a way not altogether understood, the carbon of the carbon dioxide in the atmosphere is made to combine with water to form starch ($C_6 H_{10} O_5$) which can then be either used up immediately as a food substance or may be stored up for future use in some reservoir. This process can only take place in the presence of sunlight. If a green plant be exposed to the sun for a short time, more starch will be formed than is required for immediate consumption. This may be demonstrated by staining a small portion with iodine and examining it under the microscope; the starch grains will be turned blue by the iodine, and may thus be easily distinguished.

Chlorophyll, though chiefly occurring in plants, is not restricted to them. It also occurs in some animals, where it performs the same function as it does in plants. Thus it is found in some low worms, in one of which no mouth or alimentary canal is present, and the whole of the carbon required by the worm is probably obtained from the CO_2 dissolved in the water in which the animal lives. It is also found in the cells lining the alimentary canal of the Hydra viridis. It has been lately shown that in some cases the so-called chlorophyll corpuscles are in reality Algæ, which live in a mutually helpful partnership with the animal, furnishing it with oxygen and starch, and in return obtaining nitrogenous nourishment from it. Such a partnership is described as symbiosis.

Starch occurs imbedded in the protoplasm in the cells of most plants at some period of their existence. The starch granule is a transparent body, having at one part a black spot, called the hilum which is encircled by concentric lines or striæ. It consists of two substances—starch cellulose which forms the skeleton of the grain and which is unattacked by dilute acids, and starch granulose which is readily dissolved by them. The starch

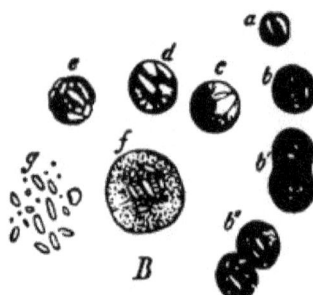

Fig. 6.—Separate chlorophyll-corpuscles with starchy contents from the leaf of Funaria hygrometrica (after Sachs). *a*, a young corpuscle; *b*, an older one; *b'* and *b"* have begun to divide; *c*, *d*, *e*, older corpuscles with starchy contents; *f* and *g*, after maceration in water by which the substance of the corpuscle has been destroyed and only the starchy contents remain.

granule grows in the same way as the chlorophyll corpuscle and the cell-wall, *i.e.* by the intussusception of new matter. Iodine colours the starch granulose blue, but the starch cellulose is coloured brown by this reagent.

The tubers of the potato and the seeds of cereals and leguminous plants are especially rich in starch granules.

Crystalloids are small masses of protoplasm, which assume a crystalline form, but possess none of the characteristics of true crystals.

Aleurone grains, or as they have also been called proteid grains, are granules of modified protoplasm, which sometimes occur alone, but very frequently surround the crystalloids.

Crystals of calcium oxalate, or calcium carbonate, occur very frequently in plant cells. They have crystallised out of the cell-sap, this being the method adopted by the protoplasm of getting rid of its excess of calcium.

Fat globules occur in considerable numbers in some seeds, and they may apparently take the place of starch in the chlorophyll corpuscles ; in these cases one must suppose that the carbon has been assimilated and fixed in the form of fat instead of starch.

HISTORICAL.

Hooke, in 1665, used the term cell, but in an indefinite way. Leeuwenhoek, in 1674, first described unicellular organisms. Fontana discovered the nucleolus in 1784 ; but it was not until Bichat, in 1801, published his "Anatomie Générale," that the idea of the whole body being made up of cells was advanced. In 1838 Schwann and Schleiden showed that most plants and animals are built up of cells, and develop from cells. Goodsir in 1845, and Virchow in 1858, proved that in every case cells arise from pre-existing cells. *Omnis cellula e cellula.*

CHAPTER IV.

CELL DIVISION.

THE division of the cell into two new ones, similar in all respects to the original one and to each other, has been studied with great care in a large number of different cells of both plants and animals. As the result of these investigations, it has been found (1) that there is a striking similarity in the mode of division in almost all cases, and (2) that the nucleus plays an essential part in the process.*

The various changes which take place in the nucleus are described as karyokinesis (κάρυον = a kernel or nucleus; κίνησις = a movement). Fig. 7 gives a drawing of nuclear division in a plant cell.

The **resting nucleus** can be seen to consist of a network of filaments which are readily affected by staining reagents, and are called chromatin threads, the remaining part, which does not easily become stained, being called the achromatin substance. When division is about to take place, the chromatin threads acquire the appearance of a coil; this is called the " **coil stage.**" The next change is that the threads divide; as they do so they take up a longitudinal direction, and appear as a series of

---

* In some cells it seems probable that a simple, so-called direct mode of division occurs. This "direct" form of division is stated to occur in some Protozoa and in some cells of higher animals. The division by karyokinesis is described as the indirect mode of division.

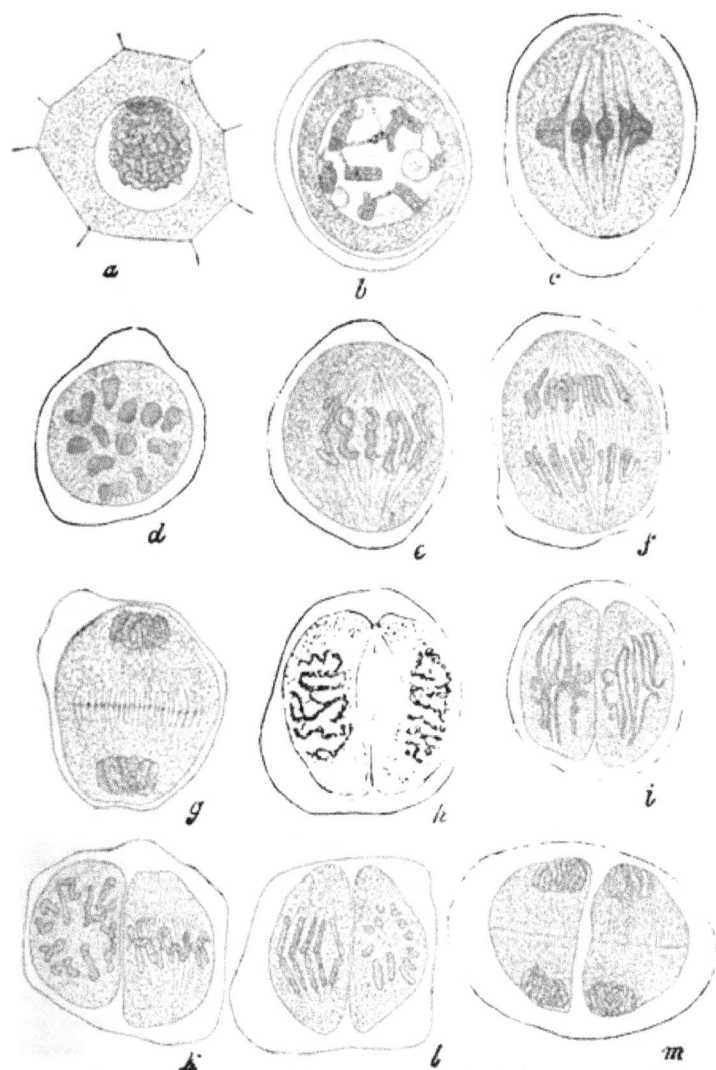

FIG. 7.—Division of the pollen mother cells of Fritillaria Persica (from Strasburger). *a*, the "coil stage"; *b*, the segments in course of longitudinal division; *c*, the nuclear spindle in profile; *d*, seen from the pole; *e*, division of the nuclear plate; *f*, separation of the daughter segments; *g*, formation of daughter coils and of the cell plate; *h*, course of the nuclear threads in the daughter nuclei; *i*, longitudinal elongation and formation of loops; *k*, nuclear spindle, to the right seen in profile, to the left from the pole; *l*, separation of the daughter segments; *m*, granddaughter coils and formation of the cell plates.

loops (fig. 7 *e*) arranged in a star-like form ; at the same time two " central corpuscles " appear in the protoplasm at opposite poles of the cell.   From these, fine achromatin threads radiate to the chromatin loops.   This is described as the **aster stage**. The next stage is characterised by the longitudinal division of the loops ; each half separates from the other, and is drawn in the direction of the central corpuscles, and hence away from the other half (fig. 7*f*).   The result of this is, that two stars are formed, slightly separated from each other (**diaster stage**). Meanwhile the cell substance has divided into two, and the threads in each star assume the form of a coil, resembling the original coil of the complete nucleus.

The division is then complete (fig. 7*g*).   Each coil now becomes a " resting nucleus," and is ready to again undergo division after an interval of rest.

To recapitulate—

1. The nucleus, to begin with, is in the **resting stage**.

2. It then takes up the coil-like form of the **coil stage**.

3. The **two central corpuscles** appear at opposite poles of the cell in the cell-protoplasm.

4. The coil stage passes into the **aster stage**.

5. Division takes place, so that the **diaster stage** is reached.

6. The cell-protoplasm divides.

7. Each nucleus again becomes a coil.

8. Each coil becomes a resting nucleus.

CHAPTER V.

REPRODUCTION AND DEVELOPMENT.

EVERY organism, however complicated, whether it be plant or animal, originally consisted of a single cell, and all the changes which go on during development, by which the young organism comes more and more to resemble the parent, are changes which started with the division of the single cell into two, the two cells thus formed into four, and so on. This division of the cell is termed segmentation, and is usually started by the process of fertilisation.

Sometimes, however, segmentation starts spontaneously, or rather as the result, either of the influence upon the cell of its surroundings, or of internal changes. The excessive growth of the cell, as in the Amœba, may produce an organism so unwieldy, that the vital processes cannot be efficiently performed. In some animals, although fertilisation is the rule, yet sometimes the egg cell develops without having been fertilised, that is parthenogenetically. In bees, for instance, whilst the fertilised eggs develop into queen bees and working bees, the unfertilised eggs give rise by parthenogenesis to drones.

In all the higher forms, and in most of the lower ones, the **germ cell or ovum** begins to **segment** as the result of a process of **fertilisation**, which will now be described.

In some animals, such as the Leech, at a very early stage in development, a certain number of cells are differentiated off for the purposes of reproduction, and it is these cells and these alone that are concerned in reproducing

23

the species, that is in giving birth to new animals similar to the parent. In such animals it is easy to make out that the reproductive cells, the so-called germ cells, are set apart before much differentiation of labour has taken place in the collection of cells forming the young animal, and that therefore these cells are very similar to the original one from which the organism has been derived. In most animals, however, including the higher vertebrates, these germ cells are not differentiated-off from the other cells until pretty late in development, and then it is not so easy to realise that they retain many of the characteristics of the original cell. In other words, the continuity of germ cells is demonstrable in some animals, but not in most. Weissman considers, however, that in all animals there is a special part of the nucleus of the egg cell, which he calls the Keimplasma, which is not used up in building up the body of the offspring, but is set apart for the purpose of

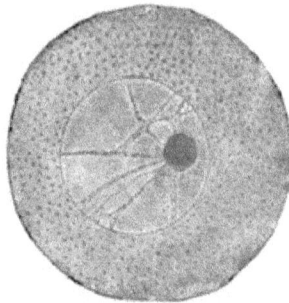

FIG. 8.—Immature egg-cell from the ovary of an Echinoderm. The large germinal vesicle (nucleus) shows a germinal spot (nucleolus), in a network of filaments, the nuclear network (after Hertwig).

forming the germ cells, which will give rise to the next generation. According to this view therefore there is not a continuity of cells throughout the history of the race, but there is a continuity of protoplasm.

The **ovum** is a typical animal cell consisting of protoplasm, a more or less well-defined cell-membrane, a nucleus (the germinal vesicle), and a nucleolus (a germinal spot).

When the ovum is fully mature it usually breaks away from the ovary, and certain changes take place in its nucleus. The nucleus moves to the surface and divides, one half remaining within the cell, but the other half being extruded. The

half nucleus within the cell again divides and pushes out
another quarter of the original nucleus, so that the egg cell
only retains a quarter of the original nucleus within its protoplasm; the remaining three quarters linger for a time
adherent to the outer surface of the ovum, but ultimately
they disappear. The extruded portions are called polar bodies,
or polar vesicles. The part of the nucleus left in the cell is
now called the **female pronucleus**.

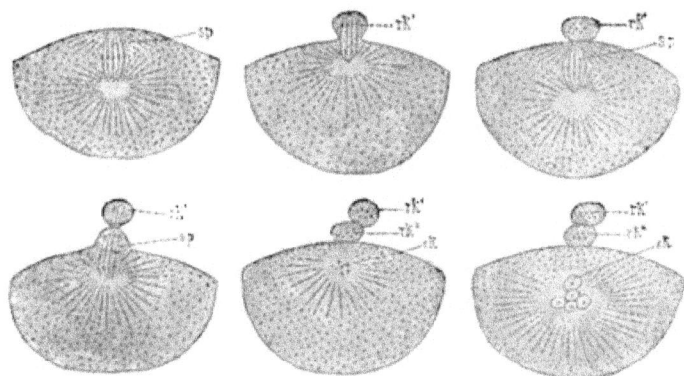

FIG. 9.—Formation of the polar bodies in Asterias glacialis. *sp,* spindle; *rk',* first
polar vesicle; *rk",* second polar vesicle; *rk,* female pronucleus (from Hertwig).

The spermatozoon or male element is also a cell, but it does
not correspond to the egg cell before it has extruded its polar
bodies.

The **sperm mother cell** from which numerous spermatozoa
arise is the homologue of the ovum.

It divides into numerous sperm daughter cells, but only a
portion of the protoplasm takes part in the division, the remainder persisting as an inert mass, around which the daughter
cells are grouped. Each daughter cell consists of a large well-
developed nucleus, around which there is a delicate covering of
protoplasm, the latter being also usually produced into a long

vibratile tail. The daughter cells, or as they are now called the
spermatozoa, drop off from the inert mass of protoplasm, and
when brought into the neighbourhood of the ovum by means
of their motile tails, they swim up to it, and one of them
succeeds in forcing its way into it.

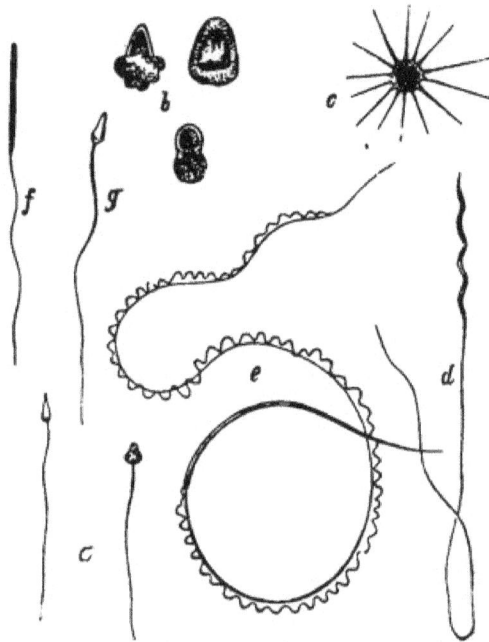

FIG. 10.—Spermatozoa (from Claus). *a*, of Medusa; *b*, of a Nematode; *c*, of a
Crab; *d*, of a Torpedo; *e*, of a Salamander; *f*, of a Frog; *g*, of a Monkey.

In some cases, the above history is modified. In these the sperm mother
cell divides repeatedly, forming a group of small cells, each of which
elongates, the nucleus taking up its position at one end, and from the other
a delicate filament growing out. The bulk of the protoplasm then collects
as a swelling upon the filament or tail, and after a time becomes completely
detached from the now fully developed spermatozoon. In both cases the
result is the same, namely that the sperm cell gets rid of a good deal of its
protoplasm before it is ready to fertilise the egg cell.

The ovum ceases to be receptive as soon as a spermatozoon has entered it, so that any others that may reach it are not allowed to get in. For a short time the head of the spermatozoon lies inside the ovum while its tail still dangles outside,

FIG. 11.—*a*, young ova of a Medusa. *b*, mother-cells of spermatozoa of a Vertebrate one of them showed Amœboid movements (from Claus).

but presently the head loses its tail and protoplasmic covering. It then consists only of nuclear substance, and is called the **male pronucleus.** The male pronucleus now approaches and eventually fuses with the female pronucleus, so that the

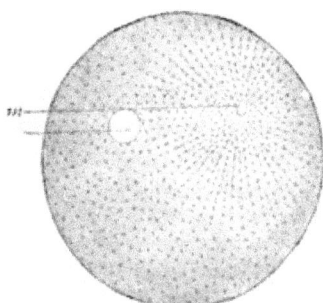

FIG. 12.   FIG. 13.

FIG. 12.—Fertilised egg of a Sea-urchin. The head of the spermatozoon, which has penetrated, has been converted into a male pronucleus (*m*) surrounded by a protoplasmic radiation, and has approached the female pronucleus (*f*) (from Hertwig).

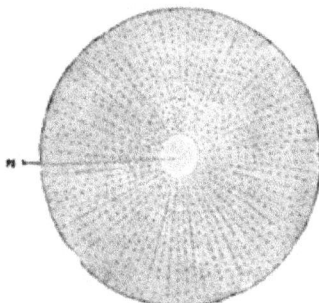

FIG. 13.—Egg of a Sea-urchin immediately after the close of fertilisation. The male and female pronuclei have united to form the nucleus (*n*) of the fertilised egg cell, which occupies the centre of a protoplasmic radiation (from Hertwig).

fertilised egg cell, with a nucleus derived from the union of the two pronuclei forms a perfect cell. It frequently contains a considerable amount of food substance or yolk.

It is interesting to notice, that the changes that take place

before fertilisation in the female cell, result in the formation
of a cell deficient in nuclear substance, whereas those which
occur in the formation of the spermatozoon, produce a cell
deficient in protoplasmic substance. As soon as the egg
cell is fertilised, segmentation begins, the single cell divides
into two cells, each of these again divides, so that four cells are
formed, the four divide into eight, the eight produce sixteen,
the sixteen thirty-two, the thirty-two sixty-four, and so on.

The appearance produced in the egg by this segmentation
varies considerably in different cases, the differences being

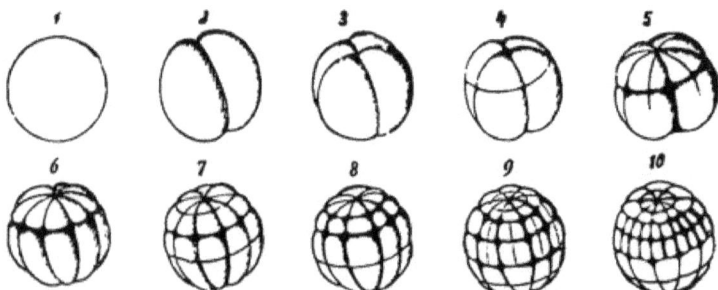

FIG. 14.—Unequal segmentation of the Frog's egg (after Ecker) in ten successive
stages.

chiefly due to the varying amounts of food substance (yolk)
present. If there be very little, the cells will be of equal size,
as in fig. 17. If there is a moderate quantity, they will be
unequal, as the food, as a rule, is chiefly collected in the
lower cells; these are generally large and unwieldy, so that
division does not take place as rapidly in them as in the
smaller upper cells.

If the food substance be present in so great a quantity that
the protoplasm of the egg cell with its nucleus is relegated to
one pole of the cell, then when segmentation takes place the
protoplasm and nucleus divide, while the inert yolk has too

little vitality to participate in this process. The result of this is, that active division only takes place over a small area at one pole of the egg, and that the numerous cells produced by this division are collected in the form of a disc or button resting upon the undivided yolk. This partial (meroblastic) segmentation, which is well seen in the Hen's egg, occurs frequently in the ova of vertebrates, occurring in Birds, Reptiles, many Fishes, and a few Mammals.

When the segmentation is complete, it is said to be holoblastic. As the result of complete (holoblastic) segmentation, a

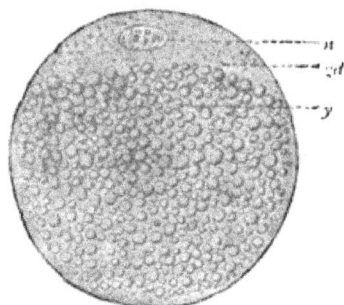

FIG. 15.—Diagram of an egg with the nutritive yolk in a polar position ; *gd,* germ disc of protoplasm ; *n,* nucleus ; *y,* nutritive yolk (after Hertwig).

mulberry-like mass of cells is formed, which is sometimes solid (**the solid morula**), but more often is hollow (**the hollow morula**). In the latter case the cells are arranged in the form of a hollow sphere (the blastula), the cavity enclosed by them being called the **segmentation or cleavage cavity.** In an egg, which contains a good deal of yolk, not sufficient to prevent complete segmentation, but enough to swell up the cells, the segmentation cavity will be comparatively small, as seen in fig. 18.

As segmentation proceeds, the upper cells divide rather more rapidly than the lower; the result of this is that the lower cells become slightly pushed into the segmentation

cavity. This position having once been taken up by the
lower cells, the more they multiply the farther do they grow
in, and the more reduced in size does the segmentation cavity
become, until at last the cells, which were at the lower pole of
the hollow sphere now come to lie underneath the upper cells.
This process of pushing or growing in is described as invagina-

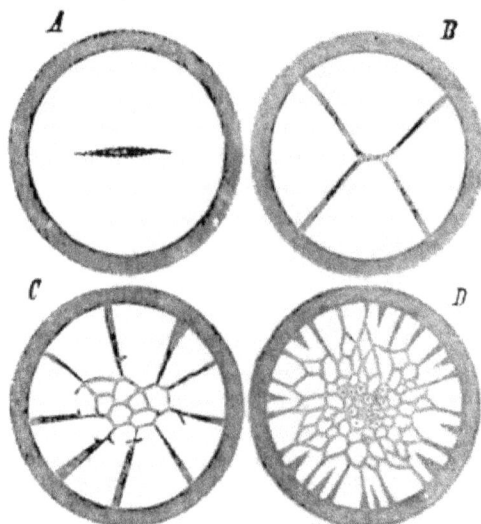

FIG. 16.—Segmentation of the germinal disc of a Hen's egg, surface view (after
Kölliker). *A*, germinal disc with the first vertical furrow ; *B*, the same with
two vertical furrows crossing one another at right angles ; *C* and *D*, more
advanced stages.

tion, and the resulting two-layered sac is described as the
gastrula or diblastula condition of the embryo. This process is
well seen in the egg of the Amphioxus, one of the lowest of
the Vertebrates (fig. 19).

The outer layer of the diblastula is called the **epiblast** or
ectoderm, the inner being termed the **hypoblast** or endoderm.
The opening, by which the cavity of the sac communicates with

the exterior, that is to say, the mouth of the sac, is called the **blastopore**, the cavity itself being described as the **archenteron**,

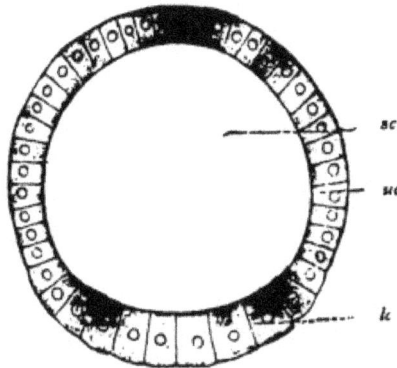

FIG. 17.—Blastula of Amphioxus (after Hatschek). *sc*, segmentation cavity; *uc*, upper cells; *lc*, lower cells.

as from it the larger part of the enteron or alimentary canal is formed.

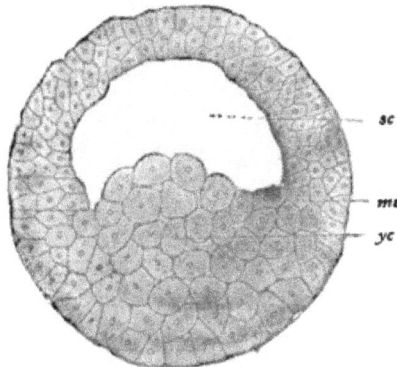

FIG. 18.—Blastula of Triton tæniatus (from Hertwig). *sc*, segmentation cavity; *ms*, marginal zone; *yc*, cells with abundant yolk.

In an egg, like that of the Chick, where the segmenting protoplasm and nucleus are confined to one pole, the resulting segmented disc is described as the **blastodermic area**, the germinal disc, or simply the blastoderm.

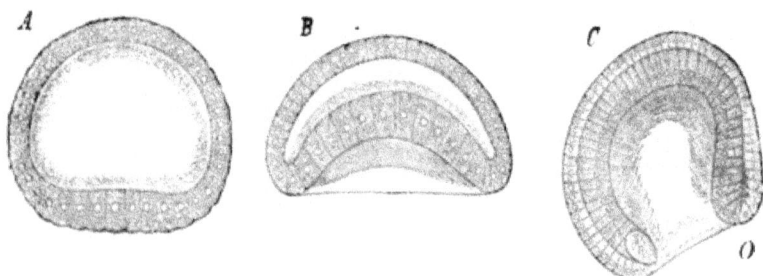

FIG. 19.—*A*, blastosphere (blastula) of Amphioxus; *B*, invagination of the same; *C*, gastrula (diblastula) invagination completed; *O*, blastopore (after Hatschek).

This blastoderm, however, is not solid, but forms a disc similar to that which would be produced by two very shallow

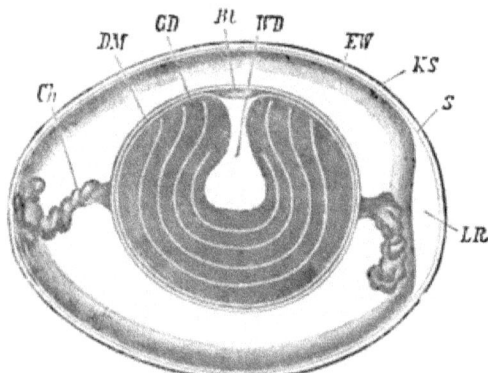

FIG. 20.—Diagrammatic longitudinal section through an undeveloped Hen's egg (after Allen Thompson). *Bl*, germinal disc; *GD*, yellow yolk; *WD*, white yolk; *DM*, vitelline membrane; *EW*, albumen; *Ch*, chalazæ; *S*, shell membrane; *KS*, Calcareous shell; *LR*, air chamber.

watch glasses, so placed as to enclose a shallow cavity. The upper layer is composed of comparatively small cells, and is continuous at its margin with the lower layer, which consists of larger ones. These are only imperfectly separated off from the yolk, on which they rest. The cavity between the two layers is the cleavage or segmentation cavity.

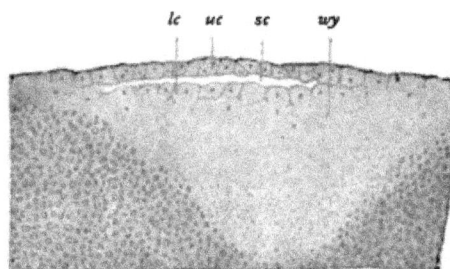

FIG. 21.—Section through the germ disc of a Hen's egg (after Duval). *sc*, segmen tation cavity; *wy*, white yolk; *lc*, lower cell layer; *uc*, upper cell layer.

As Duval has recently shown, at one margin of the blastodermic area there is a slight infolding and a growth inwards of the cells of the lower layer, the result of which is, that a cavity

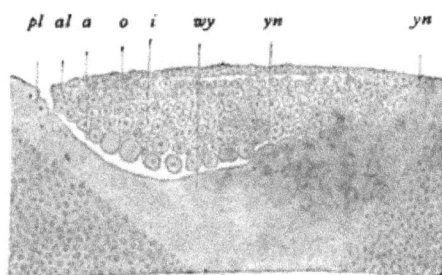

FIG. 22.—Longitudinal section through the germ-disc of a Finch (Carduelis spinus) (after Duval). *o*, outer germ layer; *i*, inner germ layer; *wy*, white yolk; *yn*, yolk nuclei; *a*, archenteron; *al*, anterior lip of blastopore; *pl*, posterior lip of blastopore.

occurs between the lower layer of cells forming the disc and the yolk. This space is the archenteron, the entrance to it being the blastopore. The cells which have grown inwards, together with those of the lower layer, are hypoblastic, and form

3

the roof of the archenteron. The hypoblast or endoderm lies immediately beneath the original upper layer, which is the epiblast or ectoderm. At a later period, the lateral margins of the hypoblast grow round to partially enclose the cavity, with the result, that the blastopore, which was at first transverse, becomes bent into what appears on a surface view to be a short longitudinal depression, and to this the name primitive streak was given, before its true meaning was understood.

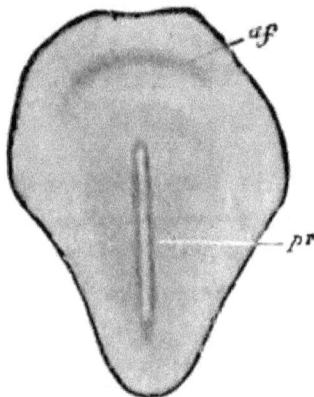

FIG. 23.—Surface view of the area pellucida in the blastoderm of a Chick, soon after the formation of the primitive groove (after Balfour). *pr*, primitive streak with primitive groove; *af*, amniotic fold.

We thus see, that in an egg loaded with food yolk, essentially the same changes take place as in the egg of the Amphioxus, and with the same result, that a gastrula is produced with its two layers (epiblast and hypoblast) and its cavity (the archenteron) opening to the exterior by a slit-like aperture (the blastopore). It will be noticed by a reference to fig. 22, that the hypoblast is more than one layer of cells in thickness, and that it does not completely surround the archenteron, the unenclosed side being composed of yolk.

During incubation, the yolk immediately under the germinal disc liquefies, the disc itself floating on the liquid, with the result that on a surface view this area appears pellucid, and is called the **area pellucida,** whilst the margin of the disc rests on the solid yolk, and has therefore an opaque appearance, being called the **area opaca.** The germinal disc slowly grows out over the yolk, but it is only the central part of it lying in the area pellucida that forms the embryo, and it is in the posterior

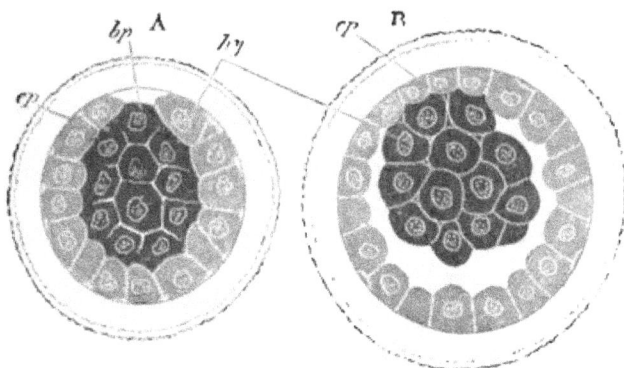

FIG. 24.—Optical sections of a Rabbit's egg in two stages immediately following segmentation (after Ed. v. Beneden). *A,* solid cell mass resulting from segmentation; *B,* development of the blastula by the formation of a cleavage cavity in the cell mass; *ep,* outer layer of cells; *hy,* central plug of cells; *bp,* place where solid plug of cells has grown into the cavity.

part of this region that the growth inwards, with the consequent formation of the archenteron and the production of the two-layered or gastrula condition, takes place.

In Mammals, the way in which the two-layered sac is formed differs somewhat from that described in the chick. The egg contains very little food yolk, hence segmentation is complete and equal, a hollow sphere of cells being formed. In the rabbit's egg a solid plug of cells is very soon seen to grow into the cavity.

This looks as though the gastrula were going to be formed, but this is not the case. A large amount of albuminous fluid collects in the remains of the cavity, with the result, that the solid plug becomes spread out at one part of the egg as a layer underneath the cells forming the sphere.

At this stage, the mammalian egg is very similar to the egg of the Chick. The albuminous fluid represents the yolk, and

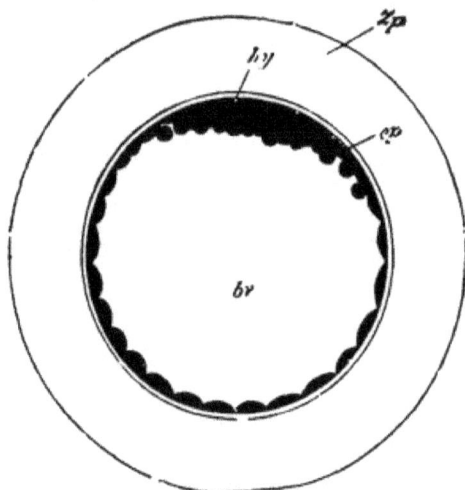

FIG. 25.—Rabbit's egg 70-90 hours after fertilisation (after Ed. v. Beneden). *bv*, cavity of the blastula; *sp*, gelatinous layer surrounding the blastula; *ep*, outer layer of cells; *hy*, spread-out central plug of cells.

the spread-out cells at the upper part, together with the cells covering them, form the blastodermic area. The chief differences between the two eggs is that the rabbit's egg is exceedingly small, only being one millimetre in diameter, and the albuminous fluid corresponding to the yolk in the hen's egg is surrounded by a layer of cells. The surrounding layer of cells has been called the epiblast, and the spread-out plug the hypoblast, but there seems to be very little doubt, that these names have been incorrectly applied, and that it is only from the mass of cells described above as the blastodermic area that the hypoblast and epiblast are formed. The way in which the further changes take place is very uncertain, but it is probable that they are similar to the changes above described as occurring in the hen's egg. Whether this be so or not, the result is the same ; that is, a germ disc is produced which shows on surface view a primitive streak, representing the blastopore, and on section an outer germ layer or epiblast and an inner germ layer or hypoblast.

We have now followed out three modes of development of the two germ layers (epiblast and hypoblast) which at first sight seem to be very different, but when compared are seen all to . consist of essentially the same processes. As the result of segmentation, a one-layered sac or blastula is produced which by the ingrowth or invagination and the multiplication of some of the cells composing it, forms a two-layered sac or gastrula, the sac being incomplete in the Chick and Mammal, but complete in Amphioxus, and in all of them having an opening at one part of it, the blastopore. The typical or normal course of events is that seen in the Amphioxus, and the course of development up to the diblastula stage, in all animals more or less closely corresponds to this type.

The next stage in development is marked by the formation from one or both of the primary layers (epiblast and hypoblast) of a third layer, **the mesoblast.** In the Amphioxus, the mesoblast is derived entirely from hypoblast, arising, as a series of segments, on either side of the embryo from the hypoblastic walls of the archenteron. Each of these segments contains within it a cavity, which enlarges to form the cœlom or body cavity.

We thus see in this case that the archenteron, or as it has

been called cœlenteron, becomes divided into two parts—a mesenteron which forms the alimentary canal, and a cœlom or body cavity.

The same arrangement occurs in other Vertebrates, only the mesoblast is given off as two solid masses of cells from the hypoblast, at first in the region of the primitive streak (blastopore), later along the roof of the archenteron. These solid masses grow downwards towards the ventral region, and later

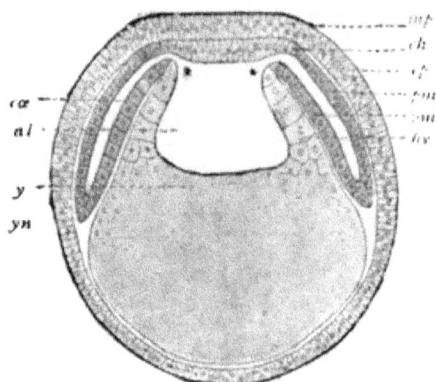

FIG. 26.—Diagram to show the development of the middle germ-layers and the body cavity in Vertebrata (from Hertwig). *mp*, medullary plate; *ch*, cells from which the notochord is developed; *ep*, epiblast; *hy*, hypoblast; *pm*, parietal layer, and *vm*, visceral layer of the mesoblast; *y*, yolk mass; *yn*, yolk nuclei; *al*, alimentary canal; *cœ*, cœlom (body cavity).

on split into two layers, one of which lines the epiblast and is called the somatopleuric layer of the mesoblast or somatopleure, while the other covers the hypoblast and is called the splanchnopleuric layer of the mesoblast or splanchnopleure.

From these three layers of the embryo, the epiblast (external layer), hypoblast (internal layer), and mesoblast (middle layer), all the complicated structures of the adult are developed.

The **epiblast** gives rise to the epidermis with its outgrowths of hairs, nails, etc., to the nervous system, to the essential parts of the sense organs, to the excretory duct of the renal organs, and to the lining of the mouth or stomatodœum, and of the posterior end of the alimentary canal or proctodœum.

The **hypoblast** gives rise to the lining of the alimentary canal, except of the entrance and exit, and to the epithelial cell-lining of the lungs, and of the digestive glands (liver, pancreas, etc.).

The **mesoblast** gives rise to all the other structures of the

FIG. 27.—Diagrammatic view of the alimentary canal of a Chick on the fourth day (after Goette). The heavy line indicates the inner germ-layer (hypoblast), the shaded portion surrounding it the splanchnic portion of the mesoblast. *lg*, lung; *st*, stomach; *p*, pancreas; *l*, liver.

body, *i.e.* the muscles, the connective tissue, the bone and cartilage, the renal organs, and probably the vascular system, and in relation with it, is developed the reproductive system.

Generally animals in their development follow the course which has been marked out in the evolution of the animals below them ; thus, a Worm at the commencement of its development is unicellular, and is similar to an Amœba, and when it has developed to the gastrula stage, it is very like a young Hydra. This law has been shortly stated in the sentence, "the development of the individual is a shortened recapitulation of the evolution of the race."

CHAPTER VI.

TISSUES.

A S we have seen, a single cell is capable of performing all the functions which are necessary for the maintenance of life, and for the reproduction of the species. The Amœba is a complete animal capable of leading an independent existence; so also Protococcus is a plant complete in itself, maintaining its own life, and giving rise to new plants to perpetuate its species. Hence, at first sight, it seems unnecessary that any animal or plant should consist of more than a single cell; but from a study of these unicellular forms, we have seen that though they are capable of doing everything necessary for the maintenance of life, yet they can only perform each function to a limited degree. The capacity of the Amœba for development is limited by the very multiplicity of the functions it has to perform; it has reached a stage where, if any further development is to take place, it must get rid of some of its functions so as to allow it to develop those that remain to it more thoroughly. If two Amœbæ were to join together, and one of them, for instance, were to limit itself to catching food, whilst the other digested it, it is evident that both Amœbæ would derive much benefit from the partnership. Again, if the separate members of a colony, composed of many Amœbæ, were to become so intimately connected together that there could be a thorough division of labour,—one group being told off to catch food for the colony, a second group to digest it, a third group to get rid of waste products, a fourth group to reproduce the species, and so on—the colony would be capable of performing the functions necessary for the maintenance of life much more thoroughly than a single Amœba possibly can. This **division of labour**, this differentiation of different groups to perform special functions, would be manifestly of the greatest advantage to the colony as a whole and also to every individual in the colony.

Every higher animal may be regarded as a colony of uni-cellular animals, and every higher plant as a colony of unicellular plants, in which this division of labour has taken place. The higher the animal or plant, the greater will be the division of labour ; that is to say, the more completely will the different cells be separated into groups with different functions.

Again, the more specialised a cell becomes, that is, the more it becomes limited to the special work for which it has been told off, the more will its shape and character differ from the cell which has to perform all functions, and the more rigid and unchangeable will it be. Further, each group will consist of a collection of cells having a like and common function, and therefore the separate cells of the group will be all, more or less, alike. Such a group of similar cells, having a common function, is described as a **tissue**.

It must not, however, be considered that an organ such as a muscle consists of only one tissue. The function of a muscle is to contract and so to approximate the parts to which its ends are attached. In order to perform this function, nearly all the cells of which it is composed have been specialised to contract when stimulated, but these actively contractile cells require to be supported and connected together by a framework of modified cells. Hence we see that a muscle must consist of two forms of tissue, the contractile tissue and the framework tissue, the latter being described as connective tissue.

In addition to these tissues, there are in themuscle, blood-vessels to supply it with nourishment, lymphatics to drain the tissues of the fluid which has escaped from the blood-vessels, and nerves to convey stimuli to the muscle and sensations away from it ; and each of these structures is made up of more than one kind of tissue.

In the same way we find that a green leaf is made up of several different tissues. The main mass of the leaf consists of chlorophyll-containing cells with large intercellular spaces between them ; the outer wall of the leaf is made up of epidermal cells, whose function is mainly protective ; and running in the leaf are fibrovascular bundles which contain various kinds of cells, some of which form a supporting tissue, others are modified to form vessels which convey water and nourishment to the leaf, and so on.

We thus see that in both plants and animals each organ is made up of several tissues, but we generally find that one of these tissues is the important one for that special organ, the rest being only present to enable the important tissue to perform its functions under the best possible conditions.

CHAPTER VII.

ANIMAL TISSUES.

STRICTLY speaking there are as many tissues in the body as there are groups of cells performing different functions, but since many of the groups, although performing different functions, have similar anatomical characteristics, all the tissues of the body are grouped together into only a few classes.

In animals the chief anatomical differences presented by the various groups of tissues from each other, depend not so much upon the characters of the cells themselves, as upon the amount and character of the substance which surrounds them and therefore separates them from each other. This substance is a product of the activity of the protoplasm of the cells, and is described as intercellular substance; if it is present in any considerable quantity, the tissue has the appearance of a matrix of intercellular substance in which the cells are imbedded.

The elementary tissues in the animal body are—

Epithelial tissue.	Bone.
Connective tissue.	Muscular tissue.
Cartilage.	Nervous tissue.

In some cases, tissues which at first sight appear to be distinct, are yet found to have so much in common in their structure and origin that they can be classed under one of the above headings.

Epithelial tissues are characterised by the small amount of

43

intercellular substance that they contain, the epithelial cells
appearing to form continuous masses. Frequently the intercellular substance is present in so small a quantity, as only to
become apparent after careful staining with nitrate of silver or
some other special reagent. Epithelial tissues occur (1) on
the surface of the skin, and on the lining of all the ducts and
recesses of the glands which open on to the surface of the
skin ; (2) on the lining membrane of the alimentary canal and
of all the ducts and glandular recesses which are connected

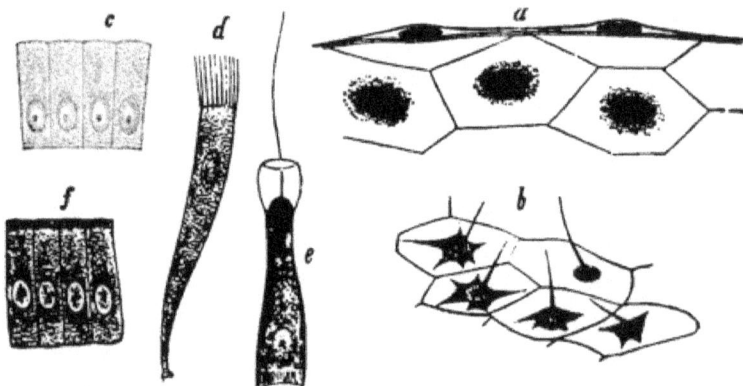

Fig. 28.—Various kinds of epithelial cells (from Claus). *a*, flat cells; *b*, flat cells
with flagella (from a Medusa); *c*, columnar cells; *d*, ciliated cell; *e*, flagellate
cell with collar (from a Sponge); *f*, columnar cells with porous borders
(intestinal epithelium).

with the alimentary canal ; (3) on the lining of the respiratory,
urinary and genital passages ; (4) on the inner or free surface
of closed cavities, such as the various portions of the body-
cavity ; (5) on the inner surface of the cavities of the heart,
blood vessels, and lymphatics.

Epithelial cells vary considerably in shape according to the
different functions they have to perform. Thus on the surface
of the skin they are flattened, whilst in the alimentary canal

they are elongated and are arranged side by side like so many columns (*columnar epithelium*). Sometimes they are "ciliated," that is to say their free surface is drawn out into numerous fine threads of protoplasm; sometimes they are arranged in a single layer; sometimes as on the skin there are several layers, forming a stratified epithelium. Hairs, nails, claws, and feathers all consist of modified epithelial tissue.

Connective tissues, Cartilage, and **Bone** may all be considered together, as they possess very many points in common. They

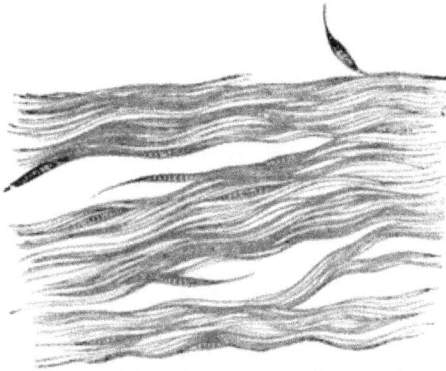

FIG. 29.—White fibrous connective tissue (from Claus).

are all developed from the middle layer of the embryo, they all serve to support or connect other tissues, and further they are capable of replacing each other in different classes of animals. Thus in the lowest Vertebrates connective tissue largely takes the place which, in Vertebrates rather higher in the scale, is occupied by cartilage, and in still higher Vertebrates by bone.

Each of these three tissues is characterised by possessing a large amount of intercellular substance forming a matrix in which the cells are imbedded. Thus in the first there is a

fibrous matrix consisting of either yellow elastic or white fibres,
the cells being called connective-tissue cells or corpuscles ; in
the second the matrix consists of a firm ground substance, the

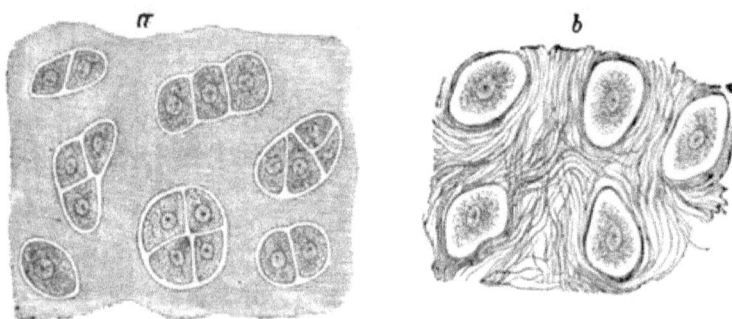

FIG. 30.—*a*, hyaline cartilage with cartilage cells ; *b*, fibro cartilage (from Claus).

cells being described as cartilage cells ; and in the third the
matrix contains inorganic lime salts, intimately connected with
a fibrous matrix, and the imbedded cells are bone cells.

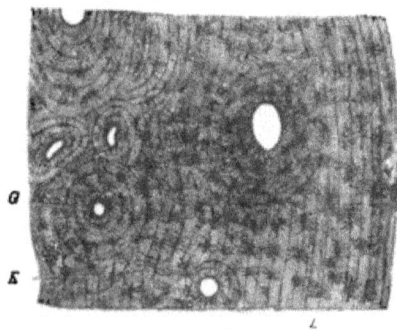

FIG. 31.—Transverse section through a long bone (after Kölliker). *K*, bone
corpuscles ; *G*, Haversian canals ; *L*, lamellæ.

Fat tissue is only a variety of connective tissue, the cells of
which have become rounded and contain large or small fat
globules. When the cell is filled with fat, the protoplasm forms

a delicate envelope to the fat, and contains the nucleus which forms a swelling at one part of the surface.

As the connective tissue is always developed from the middle layer of the embryo, and as this layer is absent in the Protozoa and in the Cœlenterata, to which groups the Amœba and the Hydra respectively belong, it is obvious that no such tissue will occur in these animals.

Amongst Invertebrates cartilage seldom occurs and bone never, the calcareous material occurring in the shells of shell-fish or on the surface of the body in crabs and lobsters, being developed on the outside of the true integument and having no farther relationship with it than that it is laid down by the integument.

Connective tissue occurs abundantly in all animals above the Cœlenterata, in most cases forming so complete a supporting framework that, could all the other tissues be removed, the animal would still retain its shape.

Muscular tissue.—The essential feature of muscular tissue is its power of contraction. In the Amœba there does not seem to be any especially differentiated part of the animal which is told off for the purposes of contraction. In the Vorticella it is the deeper layer of the ectoplasm which seems to be especially contractile. In the Hydra, contraction is brought about by the activity of the basal processes of the large ectoderm cells, which lie upon the supporting lamella. In the higher animals a special tissue is differentiated, which arises from the mesoblast, and which consists of cells modified in accordance with their functions. Two kinds of muscle cells are described —unstriped smooth or involuntary muscle, the last name being given to it because in Vertebrates such muscle is not under the control of the will; and striped or striated muscle which is characterised by having an intracellular network so arranged as to give the appearance under the microscope of alternate dark and clear transverse stripes. Each of the fibres of which the striated muscle is composed, is enclosed in a cell-membrane called the sarcolemma which has one or more nuclei lying on its inner surface.

A collection of such cells or muscle fibres as they are called, together with the supporting framework of connective tissue, constitutes one of the ordinary voluntary muscles in Vertebrates. The smooth or unstriped muscle, which usually only contracts slowly, consists of much elongated cells. The muscles of soft-bodied Invertebrates, such as Worms and Molluscs, are almost

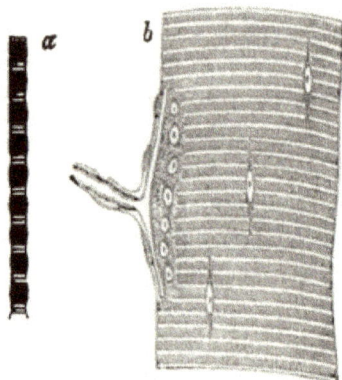

FIG. 32.—*a*, primitive fibril; *b*, striated muscle fibre of Lacerta with nuclei and nerve termination (from Claus).

always composed of non-striated cells, whereas hard-bodied Invertebrates, like the Crayfish, have striped muscle. In Vertebrates the walls of the blood vessels, of the alimentary canal, of the bladder, and of some other organs, have smooth muscle ; whereas the muscles which move the limbs, etc., and are under the control of the will, are composed of striped fibres.

Nervous tissue.—In the Amœba and the Vorticella no sign of nervous tissue occurs, the whole animal being sensitive. In the Hydra there are special ganglionic cells in the ectoderm which form the whole nervous system. In some animals of a slightly more complex structure than the Hydra, but belonging to the same group, there are superficial sensitive cells, which are connected with ganglionic cells lying below the surface, and from these ganglion cells fibres pass to the contractile elements. In the higher animals the superficial sensory cells form sense organs, the deeper cells form the ganglion cells of the central nervous system, and the fibres form the nerves.

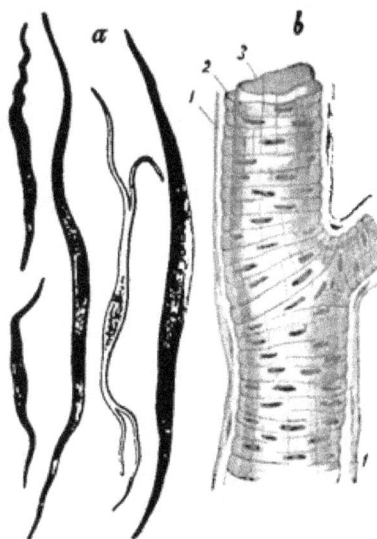

FIG. 33.—Smooth muscle (after Frey). *a*, isolated cells; *b*, piece of an artery; 1, outer connective tissue layer; 2, the middle layer formed of smooth muscle cells; 3, inner layer.

The central nervous system consists of a collection of ganglion cells and nerve fibres, which usually form a cord running the whole length of the body. In Vertebrates this cord lies in the dorsal region of the body, in Invertebrates in the ventral. In the former it is much swollen in the anterior region to form the brain. In Invertebrates the ganglion cells lie outside the fibres, whilst in Vertebrates they are collected into a mass forming the grey matter of the cord, this mass being surrounded by the fibres which form the white matter of the cord. Both cells and fibres are supported by a meshwork of connective tissue called neuroglia.

The central strand, which is the essential element of the fibre, is called the axis cylinder. Each nerve fibre is surrounded by

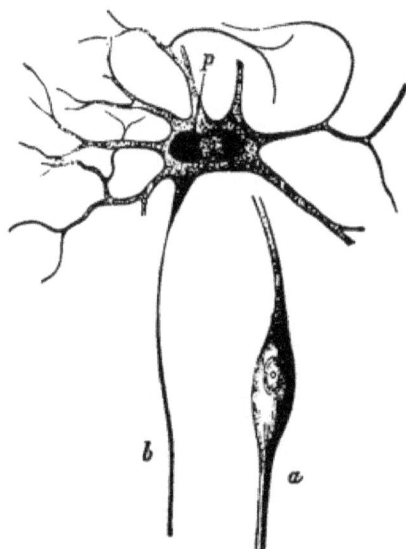

FIG. 34.—*a*, bipolar ganglion cell ; *b*, nerve cell from the anterior corner of the human spinal cord ; *p*, pigment body (after Gerlach).

a connective tissue envelope, the primitive sheath, and in addition to this, in most of the nerve fibres outside the central nervous system, there is lying between the primitive sheath and the axis cylinder, a fatty sheath, called the medullary sheath. Each ganglion cell has one or more processes. These are in direct continuation with the axis cylinders of a number of nerve fibres which correspond with the number of processes of the cell.

There are two other constituents of the body, the blood and the lymph. Both of these may fairly be described as fluid tissues, the essential cellular elements of which are represented by their amœboid cells.

The whole body is entirely composed of the above described elementary tissues, mixed in varying proportions; thus the liver and pancreas consist chiefly of epithelial tissues with only a slight admixture of connective tissue; whilst bone and muscle consist almost entirely of osseous and contractile tissues respectively. Some structures, such as blood vessels and lymphatic vessels, are composed of epithelial, connective and muscular tissues, in varying proportions. In all the larger blood vessels these are mixed in such proportions that no special tissue can be said to preponderate.

CHAPTER VIII.

PLANT TISSUES.

PLANTS are composed of tissues which are always distinguished from each other by (1) differences in the shapes of their cells, (2) the ways in which these are produced, and (3) their modes of growth.   Thus a true plant tissue is defined as a group of cells, each of which has been produced and grows in a manner similar to all the other cells of the same group. A true tissue is usually produced by one or more mother cells giving rise by repeated division to a number of daughter cells, thus forming a group.   If two true tissues coalesce, they are said to together form a false tissue.

The true tissues include in the first place growing or **meristem tissue**, in which the cells are capable of dividing ; and, secondly, **permanent tissue**, in which the cells are no longer capable of dividing.

Growing tissue is divided into primary and secondary meristem.   The former is the only tissue present in very young organs ; and the latter occurs in older organs along with permanent tissues.   All meristem cells are small, they have thin smooth walls, and are filled with protoplasm.   Permanent tissue is formed by the further growth of those cells given off by the primary and secondary meristem, which having lost the power of dividing, have assumed a permanent form.   If a thin transverse section of a comparatively young stem of a plant, such ,as a sunflower, be cut with a sharp razor, mounted in glycerine,'and examined with a low power of the microscope, it will' be at once seen that the stem is composed of several tissues. These tissues are (1) the limiting tissue or epidermis, (2) the ground, tissue, (3) the fibrovascular bundles. The **epidermis** is only one cell thick; it surrounds the other

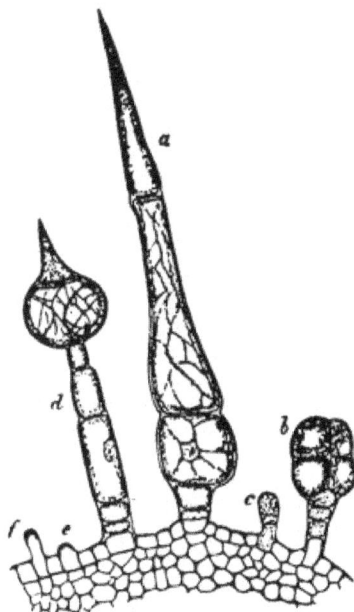

FIG. 35.—Hairs on a young ovary of Cucurbita (from Prantl). *b*, glandular hair; *c, e, f*, early stages of development.

tissues. Further it may be noticed that there are as outgrowths from it numerous hairs, each of which is composed of several cells. Each epidermal cell consists of protoplasm, a nucleus, a vacuole, and a thick cell-wall; the latter, as in all plant cells, consists of cellulose which, it will be remembered, is composed of carbon, oxygen, and hydrogen, there being no nitrogen in it, as there is in all animal cell-walls or membranes. Here and there, there are openings between the cells to allow of air passing into the interior of the stem, and the two cells guarding the opening, which are called guard-cells, usually differ from the rest of the epidermal cells in containing chlorophyll.*

The external walls of the cells are especially thick, and the outer portion can be separated as a thin cuticle, which is non-absorbent and impervious to water. Particles of wax are

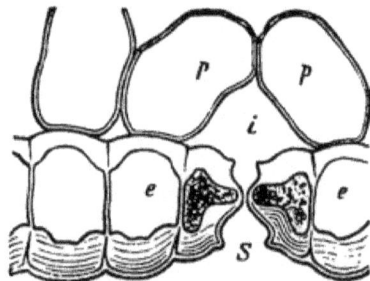

FIG. 36.—Cross section of a leaf of Hyacinthus orientalis. *e*, epidermis; *s*, stoma between the two guard cells; *i*, air cavity; *p*, parenchyma of the fundamental tissue (from Prantl).

frequently imbedded in this cuticle; these appear on the surface as fine grains, and give the bloom to plums, grapes, etc. The whole of the section within the epidermis is occupied by the ground tissue and the fibrovascular bundles. These latter will be described presently.

The ground or **fundamental tissue** contains cells which differ widely from each other in many respects, the chief differences being due to the varying thickness of the cell-walls. Thus the cells immediately in contact with the fibrovascular bundles have thick walls, and are more or less elongated

---

* In the sunflower stem all the epidermal cells contain chlorophyll; this is, however, unusual.

in a longitudinal direction. These form prosenchymatous ground tissue. Again, the cells lying in contact with the epidermis have walls which are thickened at the angles, where they come into contact with adjoining cells. These form collenchymatous ground tissue. The rest of the cells are thin walled, and these form by far the largest part of the ground tissue and constitute parenchymatous ground tissue. Many of these ground-tissue cells, especially the thin-walled par-

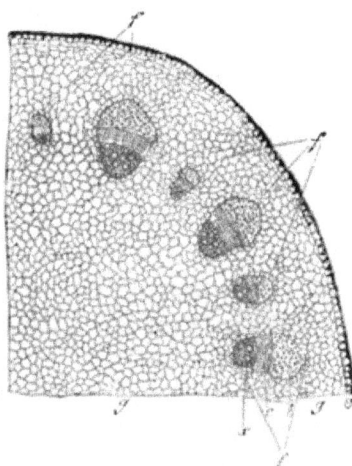

FIG. 37.—Transverse section of the petiole of Helleborus (from Prantl). *e*, epidermis; *g*, fundamental tissue; *f*, fibrovascular bundles; *x*, xylem; *c*, cambium and soft bast; *b*, bast fibres.

enchymatous cells, contain chlorophyll. As a rule, these parenchymatous ground-tissue cells do not touch each other all round, and the spaces thus left between them are described as intercellular spaces.

The **fibrovascular bundles** in the stem of a Sunflower, which may be taken as a type of Dicotyledonous plants, form a ring, dividing the ground tissue into two parts, the part outside the ring together with the epidermis forming what is known as

cortex, the part inside the ring being known as **pith**, while the ground tissue joining the pith with the cortex forms a series of rays called **medullary rays**. The fibrovascular bundles can be readily seen even with a low power of the microscope to be composed of cells differing considerably from each other in size, shape, thickness, and the structure of their walls. The part of the bundle pointing towards the epidermis is separated from the inner part by a narrow zone of small, thin-walled cells. These thin-walled cells constitute the **cambium**, and as they are capable of dividing and giving rise to new cells, the cambium is a meristem or growing tissue. If the section be examined closely, it will be seen that the cambium is not confined to the bundles, but stretches across the medullary rays from one bundle to another (this is not clearly indicated in the figure), thus forming a complete ring; the cambium within the bundle is described as fascicular cambium, and that between the bundles as interfascicular cambium. The part of the bundle outside the cambium is called **phloëm** or bast, and consists of (1) parenchymatous cells, (2) bast fibres which are elongated prosenchymatous cells, and (3) sieve tubes. If a longitudinal section be compared with a transverse one, the sieve tubes can be seen to consist of a series of cells placed end to end, the transverse walls of which are perforated plates, the sieve plates. The protoplasm of the cells forming the sieve tubes, is continuous through the sieve plates.

The part of the bundle inside the cambium is called the xylem or wood, and consists also of three kinds of cells : of parenchymatous cells, of elongated prosenchymatous wood-fibre cells, and of wood vessels. These last consist of a series of cells placed end to end, the transverse walls of which have broken down, so that a continuous tube is formed. The walls of the xylem vessels are thickened in various ways.

The above is a description of the arrangement found in the

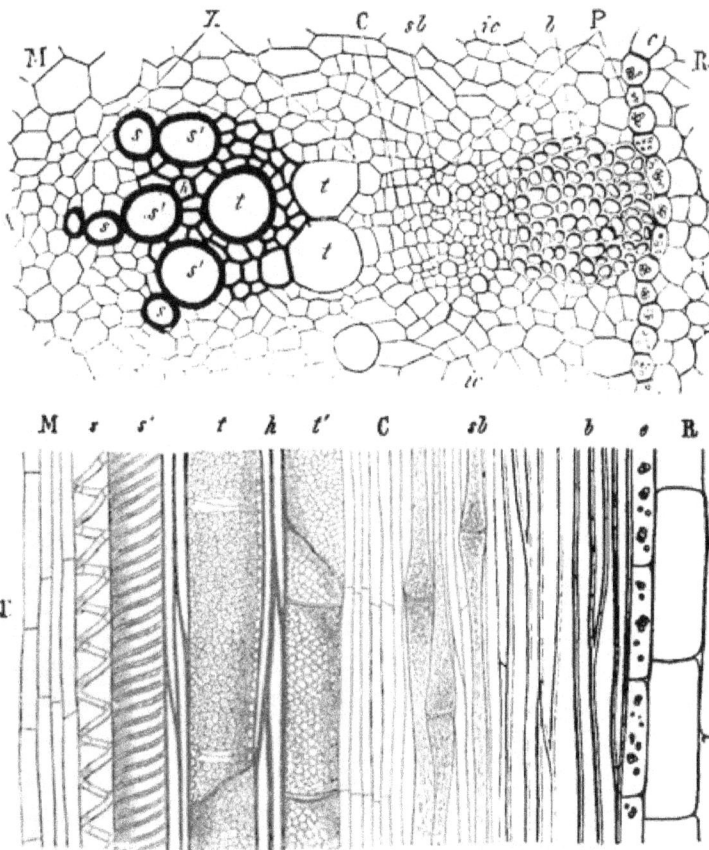

FIG. 38.--*A*, transverse section of an open fibrovascular bundle in the stem of
the Sunflower; *B*, radial vertical section through a similar bundle; *M*, pith;
X, xylem; *C*, cambium; *P*, phloëm; *R*, cortex; *s*, small, and *s'*, large spiral
vessels; *t*, pitted vessels; *t'* pitted vessels in course of formation; *h*, wood
fibres; *sb*, sieve tubes; *b*, bast fibres; *e*, bundle sheath; *ic*, interfascicular
cambium (from Prantl).

Dicotyledons and Gymnosperms. Such stems increase in
thickness by means of the activity in dividing, manifested by

the cells of the cambium-layer. The fascicular cambium, that is, the cambium within the bundle, gives off new cells externally, which develop into the various constituents of the phloëm, and internally, those which develop into the various forms of xylem cells. In the same way the interfascicular cambium gives rise to new cells which form the ground tissue of the medullary rays.

In the group of the Monocotyledons the structure of the stem is similar to that of the Dicotyledons, except that no cambium occurs; hence, no new cells can be formed. The stem, therefore, will cease to increase in thickness when the cells constituting it have attained their full size. As there is no new cell-forming layer in the bundles of Monocotyledons, they are described as definite or closed, whilst the bundles of Dicotyledons and Gymnosperms are described as indefinite or open.

In a few Monocotyledons increase in thickness is unlimited, owing to the formation of a special meristem tissue in the ground tissue; this gives rise to new closed fibrovascular bundles and new ground tissue.

As the growth of the cells in the epidermis is limited, it is obviously necessary that there should be some special arrangement made to prevent the anomaly of deeper tissues of indefinite growth being surrounded by a superficial tissue of definite growth. To obviate this condition a special meristem tissue (**the cork cambium**) forms in the ground tissue beneath the epidermis. The function of this tissue is to form internally new ground-tissue cells, which contain chlorophyll, and externally other cells whose walls undergo a special change which results in the formation of cork, a substance which, like the cuticle of the epidermis, is impervious to water. As the stem increases in thickness, the epidermis ruptures and becomes detached, the new formed cork taking its place; this again, during the following winter, ruptures, and its place is taken by the cork which has been formed during the summer.

The main part of the stem in dicotyledonous trees is formed by the permanent xylem or wood, and this wood presents a series of concentric rings, each ring marking a year's growth of the tree; this appearance is produced by the unequal size of the wood cells, according to whether they have been formed in

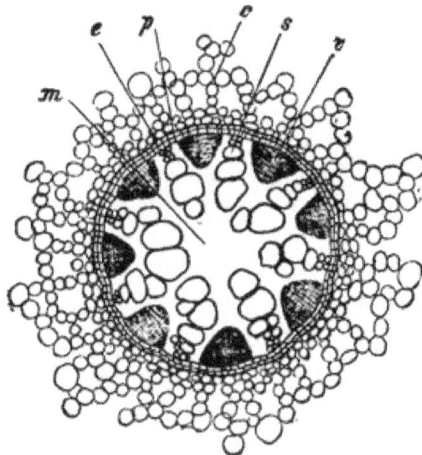

FIG. 39.—Transverse section through the root of Acorus calanus (from Strasburger). *m*, pith; *s*, wood; *v*, phloëm; *p*, pericambium; *e*, endodermis; *c*, cortex.

the spring or autumn. The cells given off from the cambium to form wood in the spring and early summer grow to a considerable size and appear light, whilst those formed in the late summer and autumn only increase in size to a comparatively small extent, and hence, being closely packed together, appear dark.

The root has the same essential structure as the stem, but the arrangement of the parts is somewhat different; thus, whereas in the stems of the higher plants the phloëm lies outside and the xylem inside, in a young root the phloëm and xylem alternate, and in addition there is a layer of meristem tissue lying outside the fibrovascular bundles, and inside the bundle sheath. This is called pericambium, and from it, in most cases, the root branches arise.

This mode of origin of root branches from the deeper tissues is characteristic of roots, they being said to arise endogenously, whereas in stems, branches arise from the superficial part of the stem, and are hence said to arise exogenously.

Van Tieghem divides the tissues of both stems and roots into three primary regions : (1) Epidermis ; (2) Cortex ; and (3) the central cylinder or stele, which includes vascular bundles, pith, etc. In most flowering plants and in the embryos of all vascular plants there is only one such cylinder ; such plants he describes as monostelic. If the cylinder branches as in most ferns, the branch he considers to correspond not to a single vascular bundle, but to a complete stele.

If a longitudinal section, made through the growing point of a stem, be examined, it will be seen that the extreme apex consists of meristem tissue, the cells of which are all almost exactly alike ; that is, at the apex, the cells have not yet become differentiated into groups of tissues.

If a longitudinal section through the growing point of the root be similarly examined, the same structure will be seen, and it will be noticed that, in addition, a cap of cells is formed on the surface of the growing point for the purpose of protecting the delicate meristem tissue from injury, as the root forces its way through the earth.

If sections be made of the stem and root at different stages of their growth, the gradual differentiation of the cells into the permanent tissues can be studied, and it will then be noticed that the cambium cells are the least differentiated, that is, in the older parts they are very similar to the meristem cells at the apex.

In many of the lower plants, such as Ferns and Mosses, the

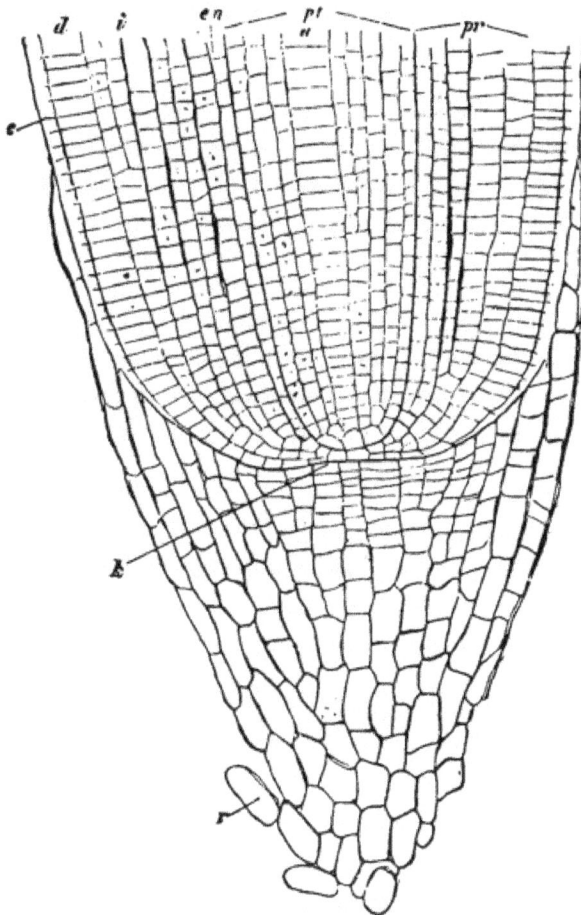

FIG. 40.—Median longitudinal section through the root-apex of Hordeum vulgare (from Strasburger). *k*, Calyptrogen; *c*, thickened outer wall of the epidermis; *d*, dermatogen; *pr*, periblem; *pl*, plerome; *en*, endodermis; *i*, intercellular passage filled with air; *a*, row of cells which will form the central duct; *r*, disorganised cells of the root cap.

extreme growing apex of the stem and root will be seen to

consist of a single pyramidal cell with its base turned outwards. From either side of this, new cells are continually being divided off. In the root the cells forming the root cap are given off from the base of the single apical cell.

Leaves have essentially a similar structure to stems, except that there is no meristem tissue at the apex. At the same time the tissues are modified in accordance with their functions; thus the epidermis, especially on the under surface of the leaf, has in it a very large number of stomata; whilst the ground tissue consists of spongy mesophyll which is composed of cells of irregular outline with very large intercellular spaces between them; the intercellular spaces are in direct communication with the external air through the stomata. The parenchyma cells on the upper surface of the leaf are very frequently somewhat elongated, and are arranged side by side in the form of a palisade. All the ground-tissue cells of the leaf contain a large number of chlorophyll corpuscles. The fibrovascular bundles in the leaves of Dicotyledons contain cambium, but this only remains active for a short time.

The leaf is of necessity limited in its growth on account of the absence of an apical meristem, and of the activity of its cambium being but temporary.

CHAPTER IX.

PLANT STRUCTURE.

PLANTS present a gradually increasing complexity of structure, as one proceeds from an examination of the lowest to the highest forms.

There are four large groups, or sub-kingdoms, the **Thallophytes**, the **Bryophytes**, the **Pteridophytes**, and the **Phanerogams**.

The **Thallophytes** include all plants in which no separation into stem structures and leaf structures can be made. The whole plant consists of a thallus, in which the tissues are only differentiated to a small extent.

The lowest members of this group are the unicellular plants, such as Gleocapsa, Protococcus, and the Yeast plant. Amongst the highest forms are the Mushrooms, and some of the more complex Sea-weeds. All Thallophytes, which contain chlorophyll, are collectively known as **Algæ**, whilst those in which no chlorophyll is found are classed as **Fungi**. The lowest forms are unicellular. In those next in simplicity, the plant consists of a colony of cells, each of which is exactly like all the others, that is, they have joined together to form a group, but there is no division of labour (Spirogyra). The next advance is, the formation of a colony of cells, in which some of the cells are told off for purposes of reproduction. Then come plants, like the Vaucheria, in which a portion of the thallus may function

63

as an organ of attachment.   A little higher in the scale, plants
are found like the moulds, Penicillium and Eurotium, in which

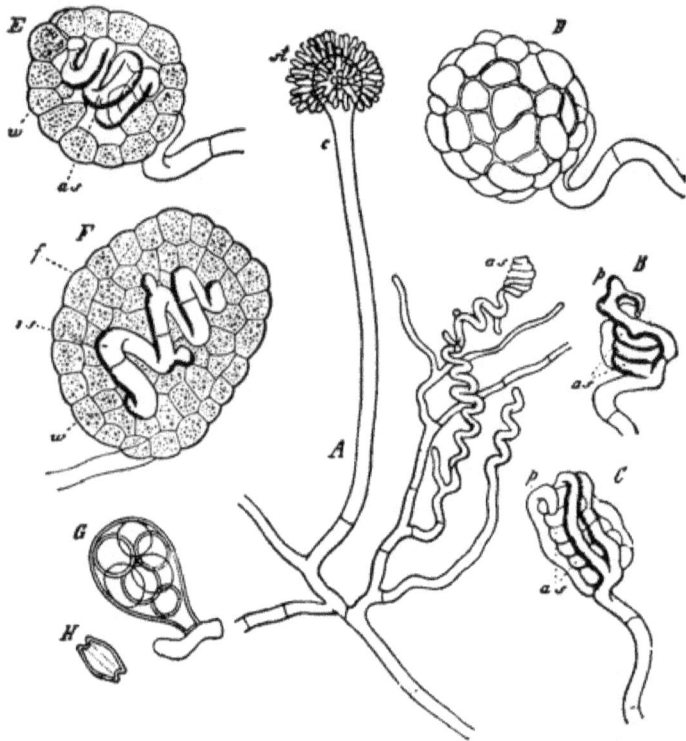

FIG. 41.—Eurotium repens (after Sachs). *A*, portion of mycelium with hyphæ ;
c, hypha bearing conidia, the conidia have already fallen off from the sterig-
mata (*st*) ; *as*, a young ascogonium ; *B*, *as*, ascogonium with a pollinodium (*p*) ;
C, another with hyphæ growing up round it ; *D*, a fructification seen on the
exterior ; *E*, *F*, sections of unripe fructifications ; *w*, the investment ; *f*, filaments arising from the ascogonium, which subsequently bear the asci ; *g*, an
ascus ; *H*, a ripe ascospore (magnified).

some cells are told off to form hyphæ or threads ; some of these
are submerged, and grow down into the soil (jam, etc.), on
which the plant is growing, others form a felt-like mass or

mycelium on the surface, whilst a third set (aerial hyphæ) are told off to bear spores (conidia). Each spore, if it fall on a suitable soil, will develop into a new thallus. In the Eurotium,

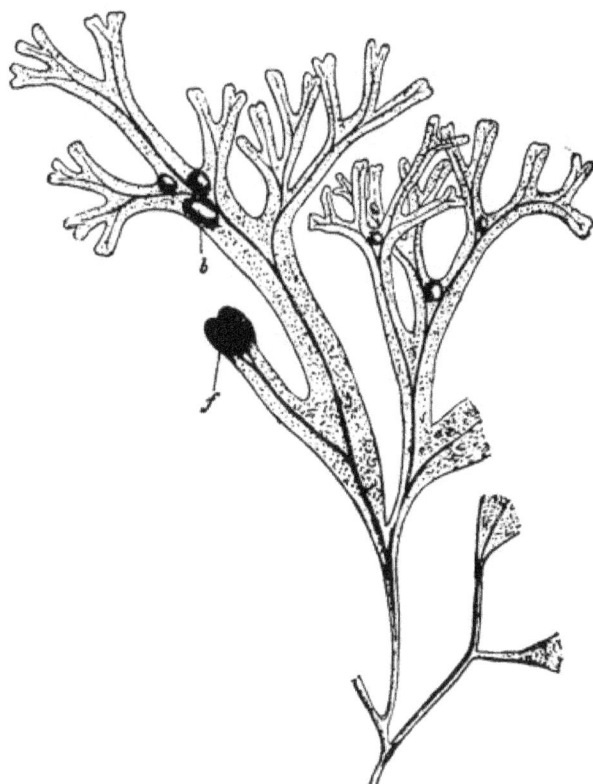

Fig. 42.—Fucus vesiculosus (Bladder wrack) (from Prantl). *b*, air bladders ; *f*, fertile branch.

in addition to this vegetative mode of reproduction, there is a sexual one, in which male and female hyphæ are produced. These unite together, become surrounded by masses of other hyphæ, and give off branches (asci) which bear spores

5

(ascospores), each of which develops into a new plant. This sexual mode of reproduction occurs occasionally also in the Penicillium.

Many Thallophytes are parasitic upon plants, and a few, such as the fungus-producing ringworm, upon animals; these are usually more complicated in their processes of reproduction. Others like the mushrooms are saprophytes, obtaining their nourishment from decomposing organic matter. In the mushrooms the whole structure appears very complicated; the spore-bearing mass (sporocarp) arises from a mycelium of threads, and bears on its under surface a number of lamellæ or plate-like structures, each of which consists of an inner cellular mass, and an outer covering of elongated cells arranged side by side, the latter giving rise to numerous outgrowths which contain spores.

In the higher Seaweeds, like the Fucus (Bladder wrack), there is an outer layer, many cells thick, forming a cortex, and an inner mass, the medulla, consisting of hyphæ. The cortex contains numerous cavities (conceptacles) which bear male and female cells, a new thallus invariably arising as the result of the fertilisation of the female by the male cells. The thallus not only presents a distinct differentiation into the cortex and the medulla, the former giving rise to modified reproductive cells, but a branching of it takes place, and, at its attached end, it is modified to form rhizoids, by which it fixes itself to stones or rocks. Many of the Seaweeds are very beautifully coloured. In many forms, as in the Fucus, the presence of chlorophyll is obscured by the existence of a brown colouring material. In some of the highest Thallophytes (Chara and Nitella) there is a still further differentiation of parts, the thallus giving rise to lateral appendages not unlike modified leaves.

A Lichen is not a simple plant, but consists of two plants, an alga and a fungus, growing together. As they are mutually helpful they form an example of symbiosis.

In the second sub-kingdom, that of the **Bryophytes** (Mosses and Liverworts), there is usually a distinct differentiation into stem and leaves, but no true roots or fibrovascular bundles occur.

The plant which bears **sexual organs**, can also be distinguished from the plant bearing only **asexual spores**, that is

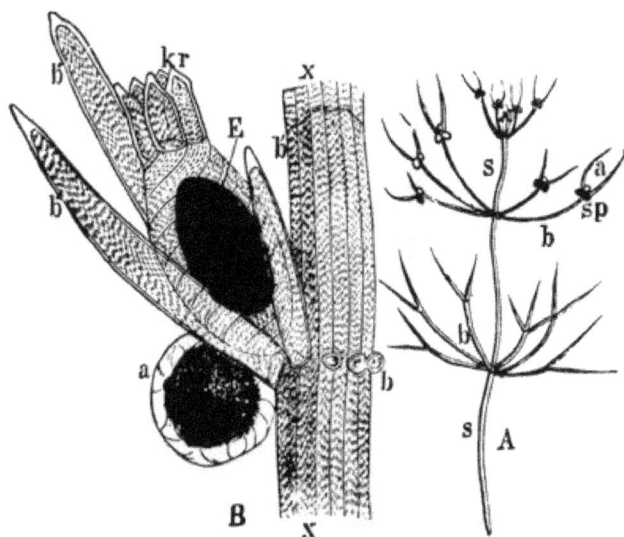

FIG. 43.—*A*, upper portion of a branch of Nitella flexilis; *s, s*, the stem; *b*, the leaves; with *sp*, the female, and *a*, the male organs; *B*, part of a fertile leaf of Chara fragilis (*x, x*); *b, b''*, the leaves; the female organ contains the oosphere, *E*; *kr*, corona of investing cells; *a*, the antheridium (after Sachs).

there are two generations. The spore grows into a sexual plant, which bears male and female organs, the male producing male elements (antherozoids) and the female producing female elements (egg cells). The egg cell is fertilised by an antherozoid, and then grows into an asexual plant, which bears only spores. Thus we have an "**alternation of generations,**" the sexual generation alternating with the asexual. The ordinary Moss plant is the sexual generation. The asexual generation grows from the fertilised egg cell, but does not lose its connection with the sexual, it being borne upon a stalk imbedded in the tissues of the sexual plant.

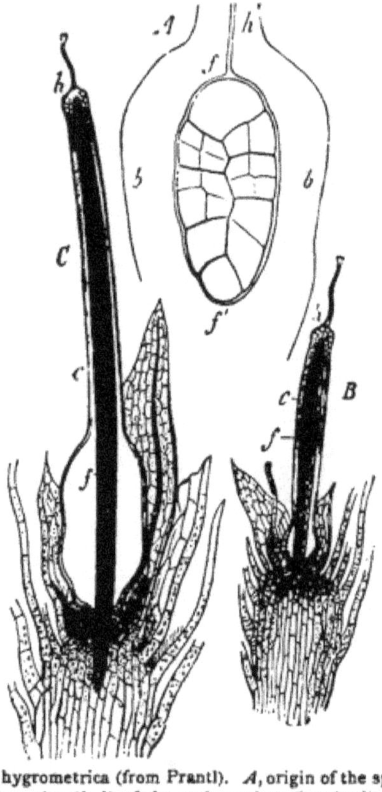

FIG. 44.—Funaria hygrometrica (from Prantl). *A*, origin of the sporogonium (*f. f*) in the ventral portion (*b, b*) of the archegonium (longitudinal section); *B, C,* different further stages of development of the sporogonium (*f*) and of the calyptra (*c*); *h*, neck of the archegonium.

Histologically a distinct differentiation may be observed. First there is an external protective layer of brownish cells, next comes a deeper layer of cells containing chlorophyll, and lastly, there is a central mass of cells, usually without chlorophyll. Running longitudinally in the central mass there is a collection of small thin-walled cells constituting the axial

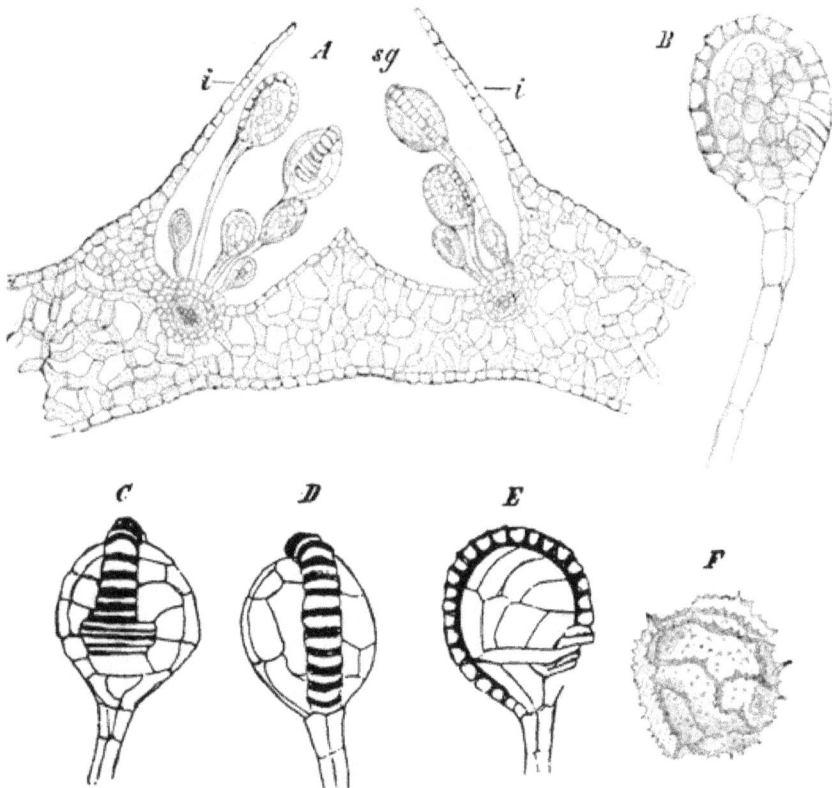

Fig. 45.—Scolopendrium vulgare (from Strasburger). *A*, transverse section through the fertile part of the leaf; *i*, indusium; *sg*, sporangium; *B—E*, sporangia; *B* and *E* seen from the side; *D* from the dorsal and *C* from the ventral aspect; *F*, a spore.

cylinder or bundle. These cells contain only water, having no living contents. Root hairs or rhizoids are developed from the lower parts of the stem; these arise from single cells, and are repeatedly branched.   In some cases there is a limiting epidermal layer of chlorophyll-containing cells, with openings in it, leading to an air chamber in which lie thread-like groups of more chlorophyll-containing cells.

The **Pteridophytes or Vascular Cryptogams** include Horse-tails, Ferns, and Club mosses.

In these plants we see that differentiation of labour has taken place to a considerable extent.   In those characters per-ceptible to the naked eye of such a plant as the Bracken Fern

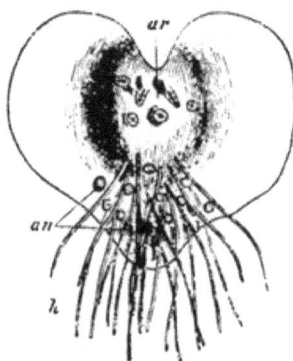

FIG. 46.—Prothallium of a Fern (from Prantl) seen from the under surface. *ar*, archegonia; *an*, antheridia; *h*, root-hairs.

(*Pteris aquilina*) we see a well-marked distinction into roots, underground stem, and branching leaves.   On the under surfaces of the leaves there are brown bodies (sori), which are collections of sporangia, each sporangium containing numerous spores.   We thus see that, whereas in the Moss the spore bearing, that is the asexual plant, is small and insignificant, in the Fern it is large.   The spores give rise to the small leaf-like bodies so constantly seen in the spring on the surface of the soil in ferneries.   Each of these is a small plant with well devoloped rootlets, and a leaf-like chlorophyll containing disc.

In most Ferns this small plant (the prothallium) bears on its under surface male and female sexual organs (antheridia and archegonia) and therefore forms the sexual generation of the Fern. The female or egg cell becomes fertilised by a male element (antherozoid), and then develops into another large spore-bearing plant, which at first retains its connection with the prothallium.

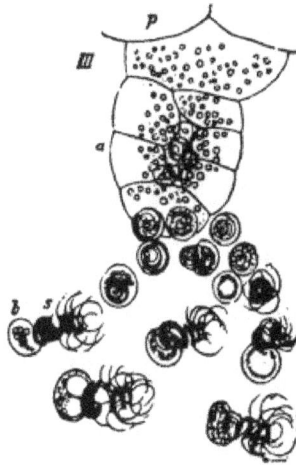

Fig. 47.—Antheridium of Adiantum Capillus Veneris (from Prantl). *p*, prothallium ; *a*, antheridium ; *s*, antherozoid ; *b*, the vesicle containing starch grains.

In some of the higher Pteridophytes, such as the Selaginella, the spore-bearing plant gives rise to two kinds of spores, small (**microspores**) and larger (**macrospores**). A microspore, when deposited on the soil, grows into a prothallium bearing only male sexual elements (antherozoids), whereas a macrospore produces a prothallium bearing only female sexual elements (egg cells) ; we thus have male and female prothallia. The egg cells of the female prothallia are fertilised by the

antherozoids of the male prothallia, and then grow up into new
spore-bearing plants.

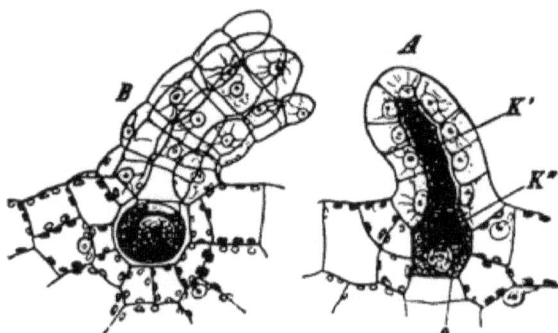

Fig. 48.—Polypodium vulgare (from Strasburger). *A*, unripe archegonium;
K', neck canal cell; *K''*, ventral canal cell; *o*, oosphere; *B*, ripe opened
archegonium.

In minute structure the Pteridophytes are far more complex
than the Bryophytes. The axial cylinder is replaced by a well-

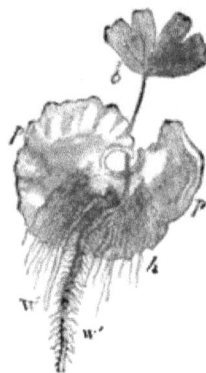

Fig. 49.—Adiantum Capillus Veneris (after Sachs). *pp*, prothallium; *b*, the first
leaf of the young fern, which is still in connection with the prothallium;
w', *w''*, its first and second roots; *h*, root hairs of the prothallium.

developed fibrovascular bundle, and the epidermis is sharply
marked off from the underlying ground tissue, which itself

becomes divided into a tissue composed of cells with thick walls (sclerenchymatous tissue) and thin-walled, chlorophyll-containing ground tissue.

In the leaves there is a well-developed spongy mesophyll, the growing point of the stem shows a well-defined single apical cell, and in the root the corresponding cell is protected by a root cap.

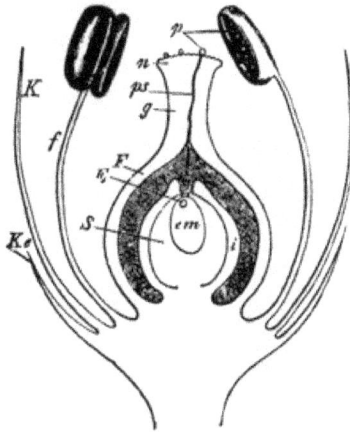

FIG. 50.—Diagram of a flower (from Prantl). *Kc*, Calyx; *K*, corolla; *f*, filament; *a*, anther with two pollen sacs open; *p*, pollen grains; *ps*, pollen tube; *g*, style; *F*, ovary; *S*, ovule; *i*, the integument of the ovule; *em*, embryo sac; *E*, the oosphere (egg cell).

The **Phanerogamia** or **Flowering plants** include all the higher plants, and are characterised by producing seeds, which are developed as the result of fertilisation, and which are borne upon the plant until they have reached maturity, when they become detached. The minute structure of the higher plants has been fully described under the head of tissues, and I would here only point out, that the differentiation of the cells to form different tissues has reached its highest development in these plants.

The **Reproductive system** in the Phanerogams consists of groups of cells borne upon and enclosed by modified leaves, the whole arrangement constituting a **flower**.

The flower stalk is much compressed longitudinally, with the result that the modified leaves forming the petals, sepals, stamens and carpels are all placed very close together on the stalk. The **sepals** are usually green, and are mainly protective in function ; they form a whorl enclosing and protecting the other parts during their development. The **petals** are usually

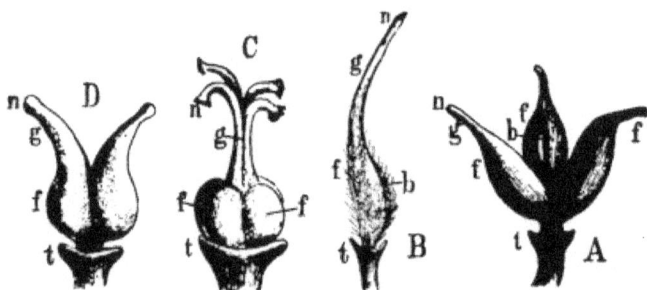

FIG. 51.—*A*, Apocarpous gynœcium of Aconite ; *B*, simple apocarpous gynœcium or pistil of Meliontus ; *C*, Tetramerous syncarpous pistil of Rhamnus ; *D*, ovary of Saxifraga, formed of two carpels ; *t*, torus ; *f*, ovaries ; *g*, style ; *n*, stigma ; *b*, ventral suture (from Prantl).

brightly coloured, and their most important function is to attract insects to the flower. The *form* of both sepals and petals is not unlike that of an ordinary green leaf. The **stamens** are much modified in form, and must be looked upon as leaves specially modified for the purpose of producing **microspores**. They are therefore described as the male organs of the plant, and are said to form the **andrœcium**. The stamen usually consists of a long thread-like stalk (the filament) on which there is a spore case or anther, generally consisting of two sacs (the pollen sacs), connected together by the con-

tinuation of the filament (the connective). Each pollen sac
or microsporangium contains within it numerous **microspores**,
or, as they are more usually called, **pollen grains**. The
carpels are leaves modified for the production and protection
of macrosporangia, usually called **ovules**; within each macrosporangium there is a macrospore or embryo sac. The apices
of the leaves are produced into a rod (the style) the top of
which is swollen to form the stigma which is provided with
a sticky secretion. Sometimes there is only one carpel, but

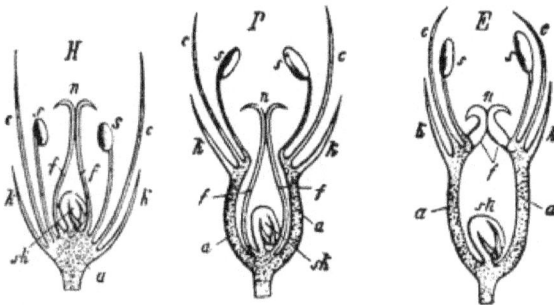

FIG. 52.—Diagram of *H* hypogynous; *P*, perigynous; *E*, epigynous flowers;
a, axis; *k*, calyx; *c*, corolla; *s*, stamens; *f*, carpels; *n*, stigma; *sk*, ovule
(from Prantl).

more usually there are three or more, and the number of
macrosporangia (ovules) present varies considerably. The
carpels form the so-called **ovary**, the whole macrospore-bearing
arrangement being described as the **gynœcium**, female organ
or pistil.

In most of the higher Flowering plants each flower contains
both stamens and carpels, but in some cases only stamen-
bearing or male flowers occur on the one plant, whilst another
produces only carpel-bearing or female flowers. In yet other
cases male and female flowers are borne upon the same plant
but on different parts of it. For example in most Gymnosperms

(Pines and Firs) the staminal leaves are borne upon one branch
and the female flowers form a cone upon another branch.
From the above description it will be seen, that the large
Flowering plant corresponds with the small asexual generation
of the Moss and with the large asexual generation of the Fern,
it (like the higher Ferns) bearing both male and female spores
(microspores and macrospores).

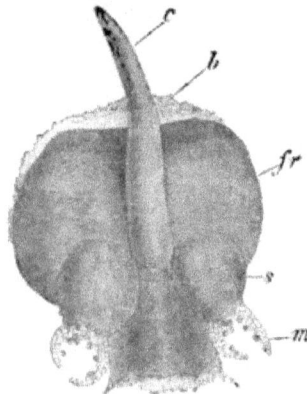

FIG. 53.—Ovuliferous scale of Pinus sylvestris (from Strasburger). *fr*, ovuliferous
scale with its two ovules (*s*) and the central rib (*c*); behind is the bract (*b*).
The integument of the ovule has grown out into two prolongations on to which
pollen grains have fallen.

Fertilisation and development.—The microspore or pollen
grain is set free by the rupture of the pollen sac, and is then
carried by the wind or by insects to another flower; if it falls
upon the stigma, it is retained there by the sticky secretion.
The subsequent processes are simpler in Gymnosperms (pines,
firs, etc.) than in the higher flowering plants (Angiosperms).
In the former the pollen grains are conducted, by means of a
rib in the scale leaf which bears the ovules, directly to the
micropyle or opening in the integuments of the ovule.   The
pollen grain then develops a tube, called the **pollen tube**, which

passes down through the tissues of the ovule until it reaches
the central part of this structure. The pollen grain and its
tube together represent a much reduced **male prothallium**.

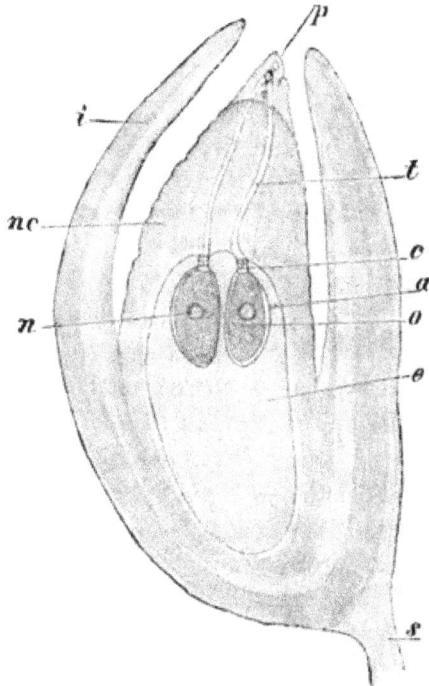

FIG. 54.—Median longitudinal section through ovule of Picea vulgaris. *e*, embryo
sac filled with endosperm; *a*, ventral portion and *c*, neck of an archegonium;
n, the nucleus of the oosphere; *nc*, the nucellus of the ovule; *p*, pollen grains;
t, pollen tubes traversing the nucellus; *i*, integument; *s*, the wing of the
seed (from Strasburger).

The nucleus of the cell from which the pollen tube has grown
divides to form two or more nuclei, which represent the
antherozoids or male sexual elements, borne upon the male
prothallium of one of the higher Ferns.

The ovule consists of integuments, and a mass of tissue, the nucellus, the central part of which forms a sac, the **embryo sac**, within which **archegonia** are formed. Each archegonium (corpusculum) bears an **egg cell**. Thus the embryo sac, in which nutrient material (endosperm) has developed, is a minute **female prothallium**, bearing female sexual elements or egg cells.

After the pollen tube has grown down and come into contact with an archegonium, one of its nuclei (antherozoids) fertilises the egg cell, which, immediately after fertilisation begins to divide, and soon develops into an embryo with seed leaves (cotyledons), a root (radicle), and a stem (plumule). The seed, containing its embryo, now becomes detached from the large plant, and if it fall on suitable soil the integuments will decay and rupture, the radicle will grow down into the soil and develop root hairs, the plumule will grow upwards and develop branches, leaves, etc., and thus a new pine tree will be produced.

In the higher flowering plants the male and female prothallia are still further reduced in size, and in detail the development is rather more complex.

The Phanerogamia are divided into two large groups : the **Gymnosperms** (naked-seeded plants, pines, firs, etc.) in which the ovule is not enclosed in an ovary, and the **Angiosperms** in which it is enclosed in an ovary.

The Angiosperms are further subdivided into **Monocotyledons** and **Dicotyledons**. In the former, which includes grasses, lilies, orchids, etc., the embryo has only one seed leaf (cotyledon), the fibrovascular bundles possess no cambium, and the leaves are parallel veined. In the latter, which includes the large majority of flowering plants, there are two seed leaves in the embryo, cambium is present in the fibrovascular bundles, and the venation of the leaves is reticulate.

CHAPTER X.

DIFFERENCES BETWEEN PLANTS AND ANIMALS.

IF one of the higher plants, such as an oak, be compared with one of the higher vertebrates, the difference between the two seems so striking that it appears by no means easy to see any points of resemblance. The plant is **stationary**; its leaves are coloured green, owing to the presence of **chlorophyll**; it is built up of cells with cell-walls, composed of **cellulose**, and its food consists of **carbon dioxide**, and **simple salts**, like sulphates, nitrates, etc., which are taken up in solution by its roots. The animal, on the other hand, **moves actively** from place to place; it does not contain any chlorophyll; its **food** chiefly consists of solid complex substances like proteids, fat, starches, etc.; and, further, its cells frequently have no definite cell-wall, and when they do the cell-wall is not composed of cellulose. Although in the main the above differences can be said to hold throughout the animal and vegetable kingdoms, yet to all of them there are numerous exceptions, so that in determining whether any given organism is a plant or an animal, one must be guided by a consideration of all its characters, placing it in the animal and vegetable kingdom according as to whether it possesses chiefly animal or vegetable characteristics.

As to the first of the plant characteristics, that is, their **stationary** mode of life, there are numerous exceptions among the lower plants. Thus the Protococcus at one stage is exceedingly active and moves freely through the water from place to place. In all other points, however, it possesses the special plant characteristics. Its possession of chlorophyll, of a cellulose cell-wall, and its mode of obtaining its food, all point conclusively to the fact that it is a plant and not an animal.

On the other hand, amongst animals, numerous instances of their remaining stationary occur. For instance, Sponges grow attached to rocks, stones, seaweed, etc.; and yet at one period of their lives they swim freely, and are evidently animals. Moreover, they feed upon complex substances, such as microscopic organisms; they do not contain chlorophyll, although in the freshwater sponge there is a green substance, which is

FIG. 55.—Growing cells of Yeast (*Saccharomyces cerevisiæ*) (from Prantl).

closely analogous to it; and, lastly, in them no cellulose occurs in the form of cell-walls.

As to the second characteristic of plants, namely, the **possession of chlorophyll**, the exceptions are also very numerous. Thus Fungi form an exception. As an example we may take the Yeast plant; it contains no chlorophyll, and therefore is unable to obtain its carbon from carbonic acid, and hence requires to be supplied with a more complex carbon-containing substance, such as sugar. Further, it is not stationary in any one place, but, on the other hand, it does not move actively, but only floats about. Moreover, it usually obtains its nitrogen, like other plants, from simple chemical compounds containing nitrogen, which are dissolved in the water in which it lives. Chlorophyll does not often occur in animals ; however occasionally it is present, as in the green Hydra (*Hydra viridis*). It is also found in some low worms.[*] In the animals in which it occurs the protoplasm is, by its agency, rendered capable of decomposing carbon dioxide, and of forming starch. In other respects, however, such as their mode of obtaining nitrogen from complex nitrogenous bodies, their active movements, the absence of cellulose, etc., these organisms have the animal characteristics, and hence are classed as animals and not vegetables.

With regard to the third plant characteristic, *i.e.*, the possession by the plant-cell of a **cellulose cell-wall**, numerous unicellular plants at some stage of their existence consist of naked protoplasm, no cellulose cell-wall being present. This is seen in the naked swarm spores of some low Algæ, such as the microzoid of Protococcus and the swarm spore of Vaucheria.

On the other hand the possession of cellulose is not confined exclusively to members of the vegetable kingdom, as a few unicellular animals possess it, and it is even found in one class of low Vertebrates, the Tunicata.

As to the fourth plant characteristic, namely, the way in which plants are nourished, several, such as yeast and other fungi, obtain their carbon from complex organic substances containing carbon, and, on the other hand, animals containing chlorophyll, obtain some of their carbon from carbon dioxide. A few plants, such as the Sundew (Drosera), obtain some of their nitrogen from complex nitrogenous compounds, such as proteids. Any small insect which alights on one of the tentacles

---

[*] In Convoluta Roscoffensis it has been conclusively shown that the chlorophyll-containing cells are Algæ, associated with the Turbellarian worm in a symbiotic partnership. It is not improbable that many of the other cases of chlorophyll-containing animals are in reality only cases of symbiosis.

of the Drosera is glued to it by means of its sticky secretion. This tentacle then bends inwards, so as to carry the insect towards the centre of the leaf, while the other tentacles bring their club-shaped ends, in which the digestive glands lie, into contact with the insect, upon which a secretion, which has the character of a digestive juice, is then poured. The proteid bodies in the insect are thus digested, or rendered soluble, and the resulting fluid is absorbed by the leaf. In all other respects these plants possess the ordinary plant characteristics.

The taking in of soluble food only is not peculiar to plants, as many parasites obtain all their nourishment by absorbing it through their body-walls.

We have thus seen that there is no single attribute of animals which is not shared by some plants; and, on the other hand, there is no plant characteristic which is possessed by plants alone; hence it is necessary to allow that plants and animals are fundamentally identical, and, in fact, are only divisions of a single vital stock.

CHAPTER XI.

INVERTEBRATA.

THE animal kingdom is divided into two large groups—the **Protozoa** which are all unicellular, and the **Metazoa** which are all multicellular. The **Metazoa** are further sub-divided into **Invertebrate Metazoa** and **Vertebrate Metazoa**. The **Vertebrate Metazoa**, which include all the vertebrated animals, will be considered in a later chapter.

The **Invertebrata**, or animals without a backbone, are characterised by the following constant features : they possess **no backbone** ; the nerve cord or nerve cords are never dorsal, but **always either ventral or lateral,** and the heart is **always placed in the dorsal region,** whereas in Vertebrata the nerve cord is dorsal and the heart ventral.

The invertebrate Metazoa are divided into the Cœlenterata and the Cœlomata.

The **Cœlenterata** have the following characteristics : there is only one cavity, the alimentary canal ; there is no body cavity ; there are only two definite layers to the body, the hypoblast and epiblast ; they have radial symmetry, and the long axis of the adult corresponds to the long axis of the gastrula (the Hydra belongs to this group). The **Cœlomata** on the other hand possess a body cavity, a distinct middle layer of cells (the mesoblast), and the longitudinal (head to tail) axis of the body does not correspond to the long axis of the gastrula.

The **Cœlomata** are again subdivided into **Vermes** (worms), a heterogeneous group of animals differing very widely from each other; **Echinodermata** (Star-fishes, Sea-urchins, etc.); **Arthropoda** (Crabs, Centipedes, Insects, Spiders, etc.), and **Mollusca** (Mussels, Snails, Cuttle-fish, etc.).

Protozoa are subdivided into three groups.

1. Rhizopoda, of which Amœba is a good example.

There are simpler forms than Amœba, and these have been grouped together under the name Protomyxa. In some of these forms no nucleus has yet been discovered.

Belonging to the Rhizopoda are a group of shell-bearing forms, most of which live in the sea; when they die they sink down to the bottom; their shells, however, persist and help to form the "ooze" which lies on the surface of the sea-bottom. Most kinds of chalk consist of shells of Rhizopoda which have accumulated in past ages; there are also enormous numbers in sea-sand, one observer having estimated that one ounce of sea-sand may contain as many as one and a half millions of shells of dead Rhizopoda.

2. Infusoria.—The Vorticella is a good example of this group.

3. Gregarinidæ (Sporozoa).—These are parasitic Protozoa, consisting of Gregarinæ (*q.v.*) and Psorosperms. The latter are parasitic upon all kinds of animals, both vertebrate and invertebrate; they have been found in the blood, are exceedingly common in the liver of the rabbit, and have been found in the human subject in the liver and in cancerous growths. The importance of these parasites with regard to disease is not yet settled.

Sponges (Porifera).—Sponges form the connecting link between unicellular animals on the one hand and multicellular animals on the other. Although multicellular, yet the individual cells, of which they are composed, are only loosely united together. Some of the Sponges, indeed, appear more like colonies of unicellular organisms than like single multicellular animals. With the exception of one fresh-water species, all sponges live in the sea, attached to rocks, shells, seaweed, etc.

The Sponge, like all other animals, starts life as a single cell ; the cell segments so form a sphere which is free swimming ; finally a two-layered sac or gastrula is formed; this attaches itself by its blastophoral end to a rock, pores form between the ectoderm cells, the free end of the animal ruptures to form an exhalent aperture, and the sponge grows up into the adult form. The endoderm cells are ciliated and loosely attached together, with pores between them. The outer layer or ectoderm consists of similar cells which, however, are not ciliated. Between these two layers is a gelatinous connecting layer, the mesoglœa, amongst the cells of which crystals of lime salts occur. Water containing food substances is sucked in by the action of the cilia of the internal cells, through the many pores present in the walls, and is passed out through the large aperture at the upper part of the sac The inner cells take up the food substances from the water as it passes through. In the adult condition of most sponges, the body becomes very complicated, owing to the folding of the endoderm to produce numerous ridges and grooves, and by the formation of numerous buds which may fuse together, or which may grow more at one part than another. As a result of the fusion of buds and unequal growth of the animal itself, a very intricate system of canals is formed. The original central canal, the depressions formed in this, the canals of the buds, and the numerous pores occurring in the body-wall all take part in the formation of this system. In the middle layers of the parent and buds, spicules of carbonate of lime are deposited in some sponges, in others spicules of silica, and in others (as the ordinary bath sponge) a framework of horny material may be developed. A single sponge may be artificially made to give rise to a considerable

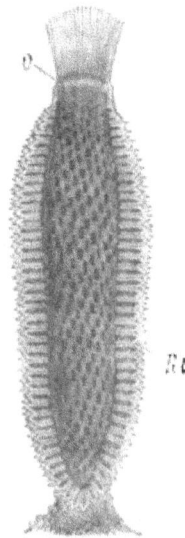

Fig. 56.—Longitudinal section through Sycon raphanus (from Claus). O, osculum with collar of spicules ; Rt, radial tubes which open into the central cavity.

number of individuals by cutting it up into pieces and placing each piece under favourable conditions for growth. This method is largely used in the propagation of bath sponges commercially. The method of reproduction in the freshwater Sponge (Spongilla) is interesting. When food becomes scarce and the water cold as in the autumn, the Sponge develops clumps of cells called gemmules ; the parent then dies, and in spring the gemmules float away and give rise to new Sponges, some of which are males and some females. Egg cells are produced in the female, and after having been fertilised by cells from the male, develop into new Sponges.

Cœlenterata.—The Cœlenterata include the Jelly-fishes, Sea-anemones, Corals, etc. The group is divided into two large classes—the **Hydrozoa,** well illustrated by the Hydra, and the **Actinozoa,** which includes the Sea-anemones. Both classes include individuals which lay down lime salts either in their tissues or outside their bodies with the result that corals are formed.

Hydrozoa.—Many of the sea forms live together in colonies. A single hydra-like form (a **polypite**) fixes itself by its foot to a rock or piece of seaweed ; and then from the foot develops a stalk which grows upwards. Upon this numerous other polypites are borne. As these polypites grow as buds from the stalk, which has itself grown as a bud from the original polypite, all the members of the colony are structurally continuous, the stomach of each being continuous through the tubular centre of the stalk with the stomachs of the rest. After a time, division of labour occurs in the colony ; some polypites become especially adapted for feeding (nutritive polypites), others are told off to function as reproductive polypites. In some free swimming forms some of the polypites are provided with air sacs to enable the colony to float, and again others with hard leaf-shaped structures which serve to protect the colony. We thus see that in the higher forms there is considerable division of labour.

In such a colony, the proliferous polypites often become much modified in form. Sometimes they detach themselves from the parent and swim about freely, being then called **medusoids.** These medusoid forms are sexual, bearing male

and female organs. After fertilisation an egg cell grows into
an embryo which, for a short period, swims about freely ; it
then settles down and grows into a new colony, which again
buds off sexual **persons**.

FIG. 57.—Branch of an Obelia stock (from Claus). *O*, mouth of a nutritive polyp
with extended tentacles ; *M*, Medusa buds on the body of a proliferous polyp ;
Th, bell-shaped cup (*theca*) of a nutritive polyp.

This life history forms a perfect example of " **Alternation of
Generations**." The asexual generation consists of the colony,
which only grows vegetatively by the production of buds.
The proliferous buds become sexual animals, and form the

sexual generation, which in its turn gives rise to a new asexual colony. The sexual persons, when they are free swimming, are rather more highly developed than the asexual forms, in that they have a nervous system, and in many cases pigment spots (so-called eyes), and a system of radial canals branching from the stomach. Some of the members of this class develop a calcareous skeleton, thus forming one kind of coral.

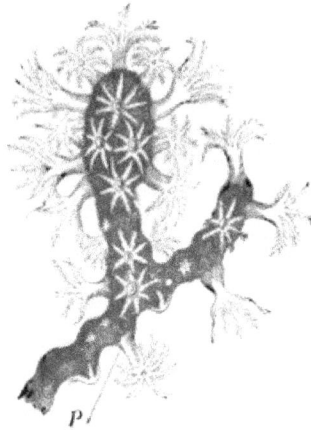

FIG. 58.—Branch of a Polyparium of Corallium rubrum (after Lacaze Duthiers). *P*, Polyp.

A jelly-fish is superficially very different from a Hydra, but in reality it is only a hydra-like animal, the body-wall of which has become thickened and widened at the base of the tentacles into an umbrella-like disc, the mouth having become elongated into a central stalk.

The second class of the Cœlenterata, which includes the Sea-anemones, is characterised by the fact that the stomach is continued into a cavity very similar to the body-cavity of the higher animals. This cavity remains, however, permanently continuous with the stomach. The cavities in the tentacles are outgrowths from it, and it is further subdivided by means of partitions (mesenteries) into a number of smaller compartments.

The Sea-anemones sometimes multiply by division, but more usually by producing egg cells and sperm cells on the margins of the mesenteries; the egg cells from the female are passed to the exterior by the mouth, they then become fertilised by the spermatozoa which have been discharged in a similar manner by the male, and then develop into new animals.

In many members of this class, coral is produced by the activity of the ectoderm cells. The organ pipe corals (*Tubipora*), the ordinary red coral (*Corallium rubrum*), the brain coral (*Mæandrina*), and the coral reefs of warm seas (*Madrepora, Astræa*, etc.) are built up by these animals.

———

Vermes (worms).—All worms are elongated soft-bodied animals, either segmented or unsegmented. In the former case the segments form a chain (Tape-worm, Leech, and Earth-worm, etc.). Most forms possess a nervous system, consisting of a mass of nerve cells (cerebral ganglia) on the dorsal surface of the pharynx, two nerve commissures surrounding the pharynx, and a single or double nerve cord extending backwards along the ventral surface of the animal underneath the alimentary canal. All worms possess a system of water-vascular tubes (nephridia). In some, as in the Leech and Earth-worm, they form segmental organs (nephridia); in others, as in the Tape-worms, they are canals extending throughout the body. Blood vessels are frequently present, as in the Earth-worm and Leech, in both of which the plasma is coloured red by the hæmoglobin, which is diffused through it. In none of these forms is there a true heart. These animals are evidently much more complex than the Cœlenterata; the three primitive layers (**epiblast, mesoblast, and hypoblast**) are all developed in the embryo, and from them are derived all the complex structures (muscles, nerves, connective tissue, glands, etc.) which are present in their bodies.

Some worms are aquatic, living in either salt or fresh water, some are terrestrial, and some are parasitic. The most important are classed as **Flat worms** (Plathelminthes), **Round worms** (Nemathelminthes), and **Ringed worms** (Annelids).

The **Flat worms** include the simple, unsegmented, ciliated Turbellarians, which live in sea or fresh water, or in damp earth ; the **Trematodes**, of which the Liver-fluke is a good example, and the **Cestodes** or Tape-worms. All the Trematodes and Cestodes are parasitic ; the former includes, in addition to the Liver-fluke, numerous species which are parasitic upon frogs and fishes, and also the **Distomum hæmatobium** or Bilharzia hæmatobia. The cercariæ of this animal may be taken in by man in drinking water. When this happens they mature quickly ; they then pass out from the stomach and make their way into the coats of the bladder and ureter, or sometimes into the kidneys, the liver, or the walls of the bowel, setting up inflammation in these organs, and giving rise to the disease known as endemic hæmaturia. The sexes are distinct, but the male carries the female about with him, she being inserted in a groove called the gynæcophoric canal.

The eggs find their way from the final host into water. They then develop into ciliated embryos, which probably make their way into the body of some small aquatic animal where they most likely grow into rediæ. These rediæ then give birth to cercariæ, which are found in large numbers in the rivers of countries like Natal and Abyssinia, where the disease known as endemic hæmaturia is common. The cercariæ may be taken in with drinking water, or, as it is supposed, they may make their way through the urethra directly to the bladder of their final host whilst he is bathing. After reaching the bladder, they bore their way into the deepe coats If they are taken into the stomach, they bore their way into the intestinal blood vessels, and then pass with the blood to the organs in which they make their final home. They then grow into adult animals, and produce large numbers of eggs, which pass out of the body of their final host with the urine. The life history of these animals has not been completely worked out, the redia stage not having been so far observed.

In some Cestodes the body consists of only one joint, with a single set of hermaphrodite reproductive organs; in others

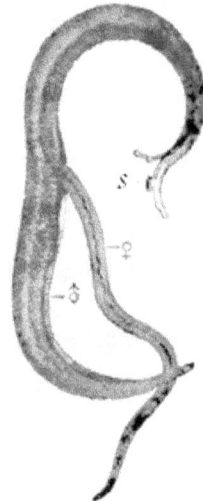

FIG. 59.—Distomum hæmatobium (*Bilharsia hæmatobia*). Male ♂ and female ♀, the latter being in the gynæcophoric canal of the former; *S*, sucker (from Claus).

there are several sets, but only incomplete formation of segments, and in yet others, as in the ordinary Tape-worm, the segmentation is well marked.

Of the next group, the **Nemertines** or Ribbon worms, the greater number live in the sea. They are unsegmented, the surface of the body is provided with cilia, and at the anterior end of the body there is a proboscis, often provided with

spines, which can be thrown out for purposes of attack or
defence. They are provided with cerebral ganglia and two

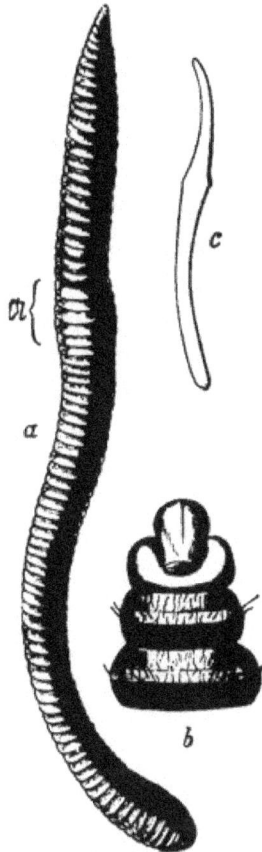

FIG. 60.—Lumbricus rubellus (after Eisen). *a*, the whole Worm; *Cl*, clitellum;
b, anterior end of the body from the ventral side; *c*, isolated seta.

lateral nerve cords, a blood vascular system, a pair of nephridia,
simple reproductive organs, and a characteristic ciliated pit,
possibly respiratory in function, on either side of the head.

The next class of worms, the **Nematodes** or Round worms, includes numerous parasites upon Man and other animals.

The **Ringed worms or Annelids** include a large number of forms which are divided into two classes, the **Chætopoda** or Bristle-bearing worms, and the **Hirudinea** or Leeches, the former being subdivided into the **Oligochæta** (or Earth-worms), the **Polychæta** which are Marine worms, and the **Echiuridæ**.

The **Earth-worm** (figs. 60 and 61), is a good example of the **Oligochæta**; it is segmented, and provided with an **alimentary canal**, consisting of a **buccal cavity, a musoular pharynx,** an **œsophagus,** a **orop,** a **gizzard,** and a **long intestine,** surrounded by so-called **hepatic cells.** There is a well-developed **blood vasoular** system, consisting of **longitudinal vessels,** which are connected together by numerous lateral branches. A well-developed **body-cavity** or cœlom is present, which has opening into it a series of **segmental organs (nephridia)**; these nephridia or renal organs open to the exterior, and form looped tubes which lie freely in the body-cavity, their internal openings forming ciliated funnels. **Respiration** is carried on by the whole surface of the body, the blood vessels coming close to the surface, and sending branches up between the bases of the epidermal cells. The whole surface of the body is sensitive to light, but there are no definite sense organs. The **nervous system** consists of **cerebral ganglia, nerve commissures** around the pharynx, and a **single nerve** cord running beneath the alimentary canal. This cord is slightly swollen in each segment to form **ganglia,** and from these ganglia nerve branches are given off to the body-walls, etc.

The **Reproductive organs.**—There are two pairs of minute testes, which are enclosed within sacs, the seminal vesicles. The **spermatozoa** undergo part of their development within these sacs, before they are passed to the exterior through the male seminal ducts (**vasa deferentia**). They are passed from one worm direct into small sacs (the **spermathecæ**, or *receptacula seminis*) of another worm, for though the worm is hermaphrodite it never fertilises itself.

There are also **two ovaries**, which drop the eggs as soon as they are ripe into the mouths of a pair of **oviducts**; they then collect in two small reservoirs attached to the oviducts

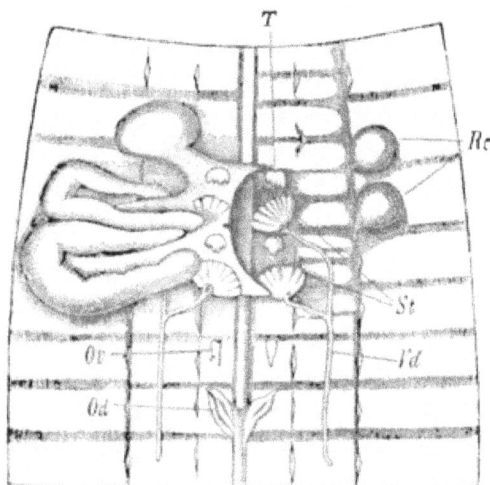

FIG. 61.—Generative organs of Lumbricus (after Hering). *T*, testes; *St*, the two funnels of the vas deferens on either side; *Vd*, vas deferens; *Ov*, ovary; *Od*, oviduct; *Rc*, receptacula seminis.

(the **receptacula ovorum**), and are finally discharged into a gelatinous substance secreted by the epidermal cells of the worm at one part of its body. At the same time the spermathecæ discharge the packets of spermatozoa (spermatophores) which they have received from another worm, and fertilisation thus takes place outside the body of the animal. Earth-worms are exceedingly numerous, Darwin having calculated that as many as 53,000 worms may occur in a single acre of ground.

They swallow earth in large quantities, extract the nourishment it contains, and then pass it out in the form of worm-casts on the surface. They are thus of the greatest importance in breaking up the soil, and so preventing the formation of a hard crust on the surface, which would eventually, but for their activity, become more or less impervious to air and moisture.

The **Polychæta**, of which the Lob-worm (*Arenicola piscatorum*) is a good example, live in the sea, and have similar habits to the Earth-worm. The Lob-worm burrows in the sand, the worm-castings, so constantly seen on sandy beaches from which the tide has receded, being produced by this animal. It is from eight inches to a foot in length ; and is provided with thirteen pairs of gills, which are thread-like processes projecting from the body and containing blood vessels. These are the respiratory organs. The Bonellia viridis, living in holes in rocks, belongs to the **Echiuridæ**; the female is of a bright green colour, and is well developed, but the male is a minute organism without either mouth or alimentary canal, and lives as a parasite upon or in the female.

The **Hirudinea** include the different species of Leeches ; they are usually small, but the Giant leech is said sometimes to measure two and a half feet in length. They mostly live in pools or marshes, but some live on dry land or in the earth. The Medicinal leech forms a good example of this class.

Allied to the Annelids are the Arrow-worms (Sagitta) and the Rotifers (Wheel animalcules). The latter are very abundant in fresh water ; they are probably degenerate animals, and are of comparatively simple structure. The appearance of a rotating wheel, from which they derive their name of "Wheel animalcules," is produced by the rapid movements of the cilia on their anterior ends.

Echinodermata.—This class includes the Sea-cucumbers, Sea-urchins, and Starfishes, all of which live in the sea.

In these animals there is an **alimentary** canal, a **body-cavity**, a simple **nervous system**, pigmented masses of cells sensitive to light (**eyes**), a blood **vascular** system, and a well-developed **water vascular system**. The water vascular system is con

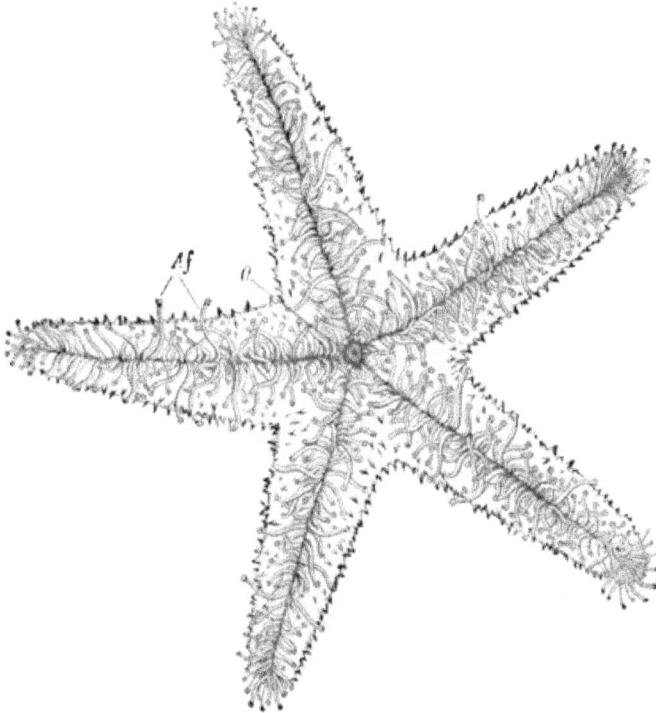

Fig. 62.—Echinaster sentus, from the oral surface (after Agassiz).
O, Mouth ; *Af,* ambulacral feet.

nected with locomotion, a condition peculiar to this class. If a Starfish be observed whilst it is moving about in the water, it will be seen that from the ventral surface of each ray of the star a very large number of small tube-like feet (**pedicelli**) are protruded. First, their terminal ends are fixed to the

7

surface of the rock on which the animal is crawling; they then
become shortened, with the result that the animal is slowly
moved on.   The tubular feet are protruded by water being
poured into them from above; then as soon as the feet are
applied to the surface of rock, etc., the water is withdrawn,

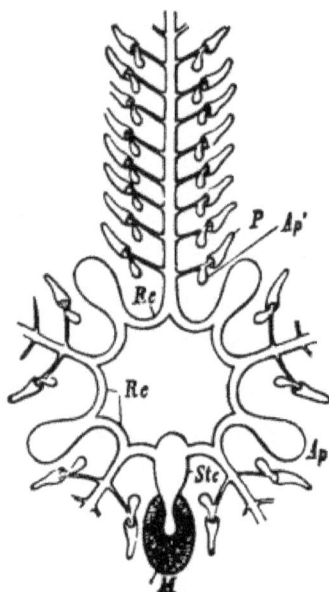

FIG. 63.—Diagrammatic representation of the water vascular system of a Starfish
(from Claus). *Rc*, circular vessel ; *Ap*, ampullæ or Polian vesicles ; *Stc*, stone
canal ; *M*, madreporic plate ; *P*, ambulacral feet connected with the branches
of the radial canals; *Ap'*, the ampullæ of the same.

with the result that the feet act like suckers and their ends
form fixed points for the animal to draw itself up to.   The
water vascular system is complicated, but briefly it consists of
the following parts : a perforated plate (madreporic plate) on the
surface of the animal, a canal with calcareous walls (stone
canal), and a circular vessel surrounding the mouth ; this

latter has in connection with it vesicles (ampullæ or Polian vesicles), and opening from it are five radial canals, which run along the arms and open into the tube feet ; at the bases of the tube feet there are small swellings or ampullæ into which the water, that is returned from the tube feet, flows.

Calcareous skeleton.—In the deeper layers of the skin in most forms lime salts are deposited, giving to the animal a definite shape, and providing it with a very complicated exoskeleton of plates and spines. In connection with the exoskeleton of the Starfish and Sea-urchin, on the dorsal surface of the former and around the mouth of the latter, there are modified movable spines, the pedicellariæ, which are of use in seizing food. The **muscular system** is only slightly developed in most Echinoderms. In the Sea-cucumbers, however (Holothuroidea), in which the calcareous plates of the skin are small and insignificant, and which are therefore rather soft-bodied, and in the brittle stars (Ophiuroidea) where the water vascular system has no locomotor function, the muscular system is well developed.

The **Nervous system** is simple, consisting of a ring around the mouth with nerves extending from it along the arms, and a superficial network of fibres and cells lying beneath the ectoderm. Nerves are also given off to the feet to the "eyes," when present, and in the Sea-cucumbers to the tentacles around the mouth. In the Crinoidea (Feather-stars) the nervous system is double.

The **Alimentary system** is also simple. In some cases it ends blindly, but generally there is both mouth and anus. The mouth is ventral, and the anus dorsal in the Starfishes and Sea-urchins. In the Brittle-star there is no anus, and in the Feather-star the mouth and anus are close together. In the Starfishes prolongations from the stomach extend into the arms, each of which has on it two pouches provided with digestive glands.   In the Sea-urchins the mouth is provided with a complicated masticating apparatus (Aristotle's lantern) which serves to crush the food.   In the Sea-cucumbers the posterior portion of the intestine is dilated, and gives off a pair of much-branched organs, the so-called "respiratory trees."   In the Feather-stars the food canal is ciliated.

The **Blood vascular system** in the Sea-cucumbers is very indefinite, consisting of spaces only.   In the other forms there are one or more vascular rings, from which branches pass to the various organs of the body, and also a collection of blood vessels, the plexiform organ, by the side of the stone canal.

The **Respiratory system.**—Starfishes and Sea-urchins breathe by means of outgrowths from the skin which hang freely in the water, and into which the body-cavity extends.   The fluid in the body-cavity contains brown pigmented cells, which seem to be capable of absorbing oxygen.   As the body is so continually being washed through by the water, it is not improbable that it partly obtains its oxygen from the water which passes through the water vascular system.

Excretory system.—The water vascular system probably functions as an excretory system.

Reproductive system.—The sexes are usually separate, the ovaries and testes resemble each other in external appearance, only differing in their products ; they open by ducts to the exterior.   The fertilised egg cell develops into a free swimming larva which does not grow directly into the adult.   The adult grows as it were inside the larva, appropriating to itself such of the larval structures as will be of use to it in its adult condition, and casting off the rest.   During this process the bilateral symmetry of the larva becomes changed into the radial symmetry of the adult.

Many Echinoderms possess, to a marked degree, the power of reproducing parts which have become detached. Thus, a Starfish will reproduce a lost arm; and a Sea-cucumber, if

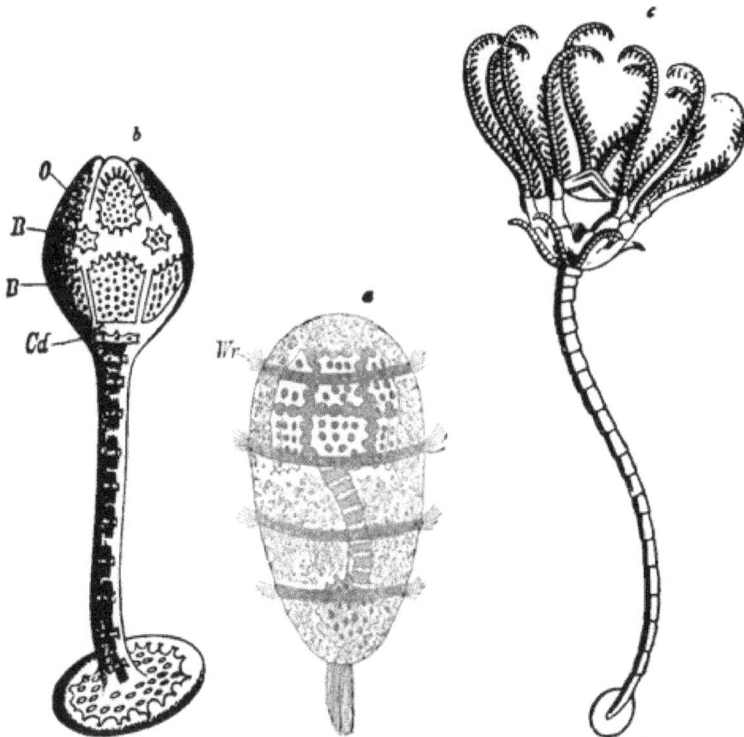

Fig. 64.—Developmental stages of Comatula (*Antedon*) (after Thomson). *a*, free swimming larva with tuft and rings of cilia (*Wr*), also with rudimentary calcareous plates; *b*, attached Pentacrinoid form of the same animal; *c*, older stage.

attacked, may eject its viscera, escape, and grow new viscera again.

The Starfishes live for the most part at the bottom of the sea, and feed chiefly upon small animals; the Sea-urchins, on the other hand, feed chiefly upon seaweeds, which they crush by means of their masticatory apparatus. The Sea-cucumbers move about chiefly by contracting their muscular bodies ; they live for the most part upon small animals which in some cases they catch with their waving tentacles.

The Brittle-stars move about in a wriggling fashion by contracting their muscular arms. Most of the Feather-stars, on the other hand, are fixed either permanently or temporarily by a jointed stalk, while others are free swimming.

***

Arthropoda are bilaterally symmetrical animals with jointed appendages and segmented bodies. They never possess cilia.

Five classes are distinguished—

1. **Crustacea**, including Crabs, Lobsters, Water-fleas.
2. **Protracheata.**
3. **Myriapoda**, including Centipedes and Millipedes.
4. **Insecta**, including Ants, Butterflies, Flies, Beetles, etc.
5. **Arachnoidea**, including Spiders, Mites, and Scorpions

Crustaceans.—This large class includes such diverse animals as Crabs, Lobsters, Shrimps, Water-fleas, Woodlice, and Barnacles. They resemble each other in having **jointed appendages**, well-developed **sensory organs, eyes,** a more or less distinctly **segmented body, a simple alimentary canal,** and a **horny exoskeleton,** which is derived from the ectoderm, and which is generally more or less calcified. Some forms have auditory and olfactory organs.

In addition to the active well-developed forms there are over seven hundred species of parasitic Crustaceans ; these, as is general with parasites, are more or less degenerate, sometimes consisting, as in Sacculina, of hardly anything but a sac of reproductive organs. In its embryonic condition this animal is similar to the embryo of many other Crustaceans having well-developed appendages, and swimming actively about.

Most Crustaceans live in water, but a few, such as Woodlice, live on dry land. In most cases the body can be divided into a **head** with five pairs of **appendages**, a **thorax** bearing appendages, and an **abdomen** with or without appendages. The

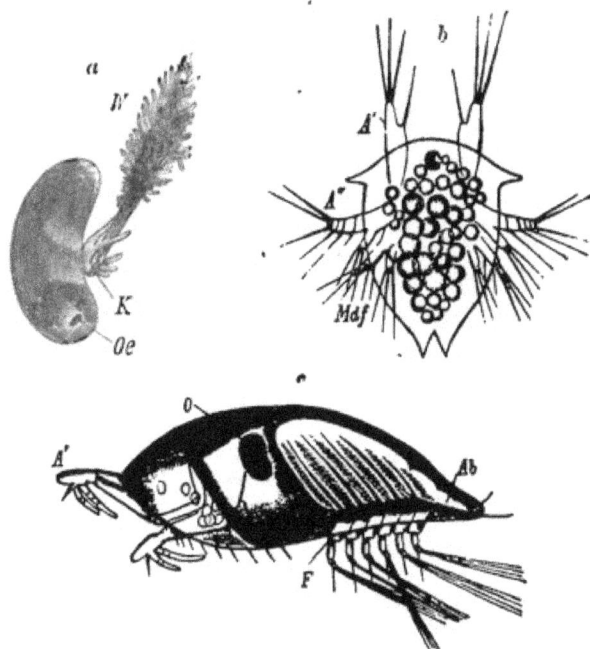

FIG. 65.—*a*, Sacculina purpurea (after Müller); *Oe*, aperture of the mantle sac *W*, root-like processes; *K*, genital aperture. *b*, Nauplius larva of Sacculina; *A'*, *A''*, *Mdf*, appendages. *c*, pupa of Lernæodiscus porcellanæ (after Müller); *F*, the six pairs of legs; *Ab*, abdomen; *A'*, attaching antennæ; *O*, eye.

calcareous covering of the anterior region, including head and thorax (**cephalo-thorax**), is frequently unsegmented, whereas that of the abdomen is usually divided into **segments**, which are movable on each other. When the abdomen bears appendages these are not infrequently modified for swimming purposes. The **blood vascular system** is but ill developed in the lower forms ; in the higher forms there is, however, a heart, arteries, and venous sinuses. In the Crayfish, for instance, the dorsal **heart** pumps blood through the **arteries** to the different organs and tissues ; the blood is re-collected from these into **sinuses**, and through them passes to **vessels in the gills.** Here it is aerated, and then sent on to a space (**the pericardial sinus**) in which the heart lies. From the pericardial sinus it

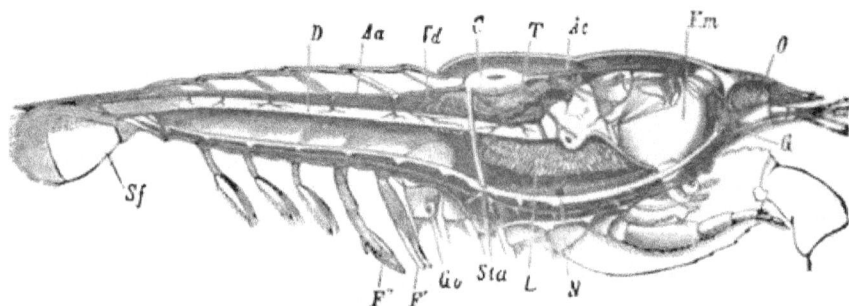

FIG. 66.—Longitudinal section through the Crayfish (Astacus fluviatilis) (after Huxley). *C*, heart ; *Ac*, cephalic aorta ; *Aa*, abdominal aorta, the sternal artery (*Sta*) is given off close to its origin ; *Km*, gastric mill ; *D*, intestine ; *L*, liver ; *T*, testis ; *Vd*, vas deferens ; *Gö*, genital opening ; *G*, brain ; *N*, nerve cord ; *Sf*, appendage of telson.

passes into the heart through valvular apertures in the cardiac wall.

The **Respiratory organs** vary considerably. In some of the lower forms aeration of the blood takes place as it courses under the skin ; in others, water passes in and out of the anus, forming a condition of **anal respiration** ; in the higher forms there are usually definite **gill filaments.**

Most Crustaceans are active in their movements, but a few (Barnacles, Acorn-shells, etc.) are fixed upon logs of wood, rocks, etc., and many of the degenerate parasites lead a very passive adult life.

Many Crustaceans conceal themselves in a growth of seaweed, sponge, etc., in the tunic of an animal like the Sea-squirt or in the shell of some mollusc (Hermit-crab in shell of a Gasteropod).

In other species the real form is often obscured by the development of shields (Apus) and shells (Water-fleas, Barnacles).

The development of the Crustacea is very interesting. In most forms a series of metamorphoses take place after the animal quits the egg. Some, like the *Penæus*, a species of shrimp, have a very complicated life history. After fertilisation and segmentation, the embryo develops into a **Nauplius** which, after a short period, **moults** and develops into a **Zooea**, a very different looking animal from the Nauplius. After a brief interval a further moulting and development takes place, with the result that the **Mysis** stage is reached; this again has a last moult, and the adult **Penæus** is produced. The Lobster starts life at the Mysis stage; and the Crayfish is born as a Crayfish, only differing from its parent in size.

The lower Crustaceans (*Entomostraca*) usually start life in the Nauplius form, but stop short soon after, the adult not being very unlike the Nauplius; while the higher forms (*Malacostraca*) do not, as a rule, pass through the Nauplius stage at all, but quit the egg at a higher level, such as the Zooea stage (Crabs), or the Mysis stage (Lobsters).

The remaining Arthropods, namely, the Protracheata, Myriapoda, Insecta, and Arachnoidea, are included under the common name of **Tracheata**.

Most of the Tracheata live either on land or in the air; but in many cases, part, and in a few, the whole of the life is spent in the water. **Respiration** is carried on by a system of air tubes or **tracheæ**, which are usually distributed through the body; but sometimes, as in scorpions, they are collected in special chambers (**pulmonary sacs**). Spiders have both pulmonary sacs (*fantracheæ*, or lungs) and ramified tracheæ.

The **Protracheata** are represented by the single genus **Peripatus**. The Peripatus is a worm-like animal, possessing tracheæ, a series of excretory tubes or nephridia, a soft body, stump-like legs, a well-developed muscular system, which consists chiefly of unstriped fibres (whereas in most Arthopods the muscles are striated), a dorsal brain, and two lateral-ventral nerve cords, an alimentary canal into which numerous glands open, and a simple circulatory system.

It will be seen by the above description that Peripatus forms a sort of connecting link between ringed worms (Annelids) and Tracheate Arthopoda.

FIG. 67.—Gryllus campestris ♂ (règne animal).

Myriapoda (Centipedes and Millipedes) resemble worms in form, but they are provided with numerous **jointed appendages**. In other points their structure resembles that of the next class (Insecta). The Centipedes are poisonous and carnivorous, whilst the Millipedes are harmless vegetable feeders.

Insecta.—This class comprises an enormous number of species; it has been stated that there are about two million species of living animals, and that nearly half of these belong to the Insecta. The majority have **wings** and the power of true flight, and are provided with a large number of **air spaces** or **tubes** (tracheæ) within their bodies. They are mostly very

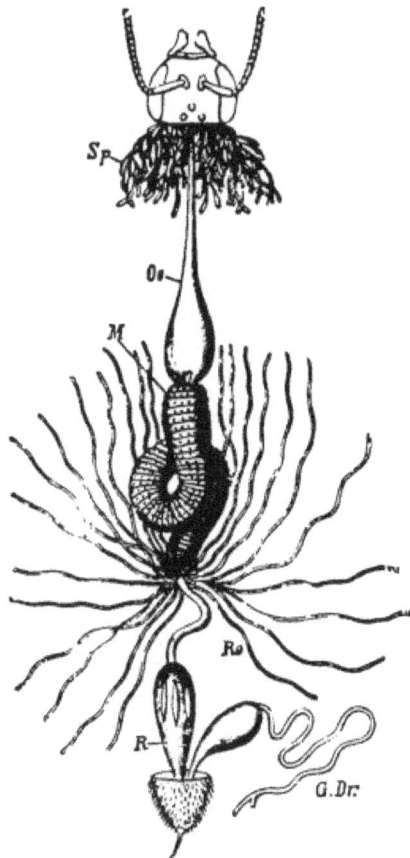

FIG. 68.—Digestive apparatus of Apis mellifica (Honey bee) (after Léon Dufour). *Sp*, salivary glands; *Oe*, œsophagus with crop like dilatation; *M*, chylific ventricle; *Re*, Malpighian tubes; *R*, rectum with rectal glands; *G.Dr*, poison glands.

active and brightly coloured, and have well-developed **nervous systems** and **sensory organs**. Their bodies are segmented, and are divided into the **head**, which bears a pair of outgrowths, the antennæ, and three pairs of **appendages**; the **thorax**, divided into **pro-**, **meso-**, and **meta-thorax**, each bearing a pair of legs; and the **abdomen**, usually destitute of appendages. The appendages are all jointed. The wings are borne upon the meso- and meta-thorax, and are flattened sacs, moved by muscles, and possessing veins or nervures, which consist of tracheæ, nerves, and outgrowths from the body-cavity. Many Insects, however, like Lice and Fleas, do not possess wings. The movements of the wings are usually very rapid; those of the Fly, for instance, give about 330 strokes per second. The **skin** forms a chitinous investment to the body. The **muscles** are of the striated variety. The **cerebral ganglia**, the double ventral nerve chain of ganglia, and the **sense organs** are all well developed; the **eyes** are made up of a number of elements, each one consisting of a corneal facet, a lens-like structure, and a nerve end organ. Sounds are produced by the rapidly moving wings, as in Flies; by the legs scraping against the body, as in Grasshoppers, and in some other forms by complicated special structures. Their food is very various, comprising leaves, honey, seeds, blood, wood, and flesh. The alimentary system is usually made up of a **mouth-cavity**, with **salivary** glands, a **pharynx**, an **œsophagus**, a **crop** or **gizzard**, a **mid-gut** (**mesenteron**), with **digestive glands** opening into it, and a **hind-gut**. The **respiratory** system consists of branching air tubes, the air being pumped in and out by the panting movements of the animal.

In the **larvæ** which live in water, the tracheal system is closed, and they breathe during their larval state, either by gills, containing tracheæ, which grow out from the sides of the abdomen, or through the anus (anal respiration).

The **Circulatory** system is simple, for owing to the ease with

which oxygen can get to the tissues through the branched
tracheæ, it is unnecessary that the blood should function as an

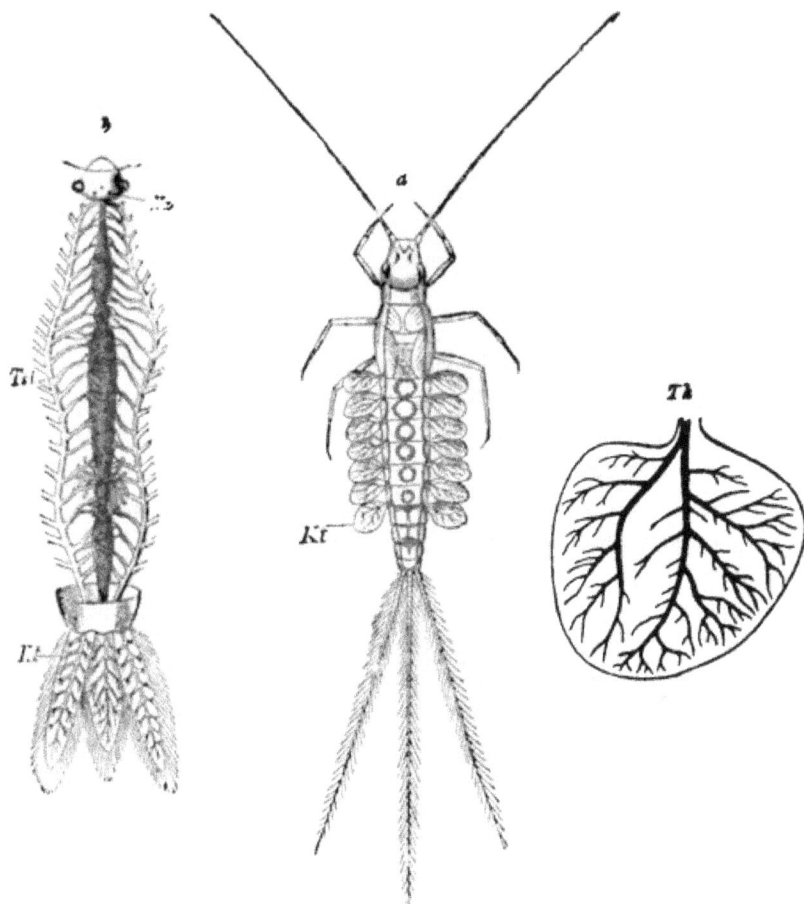

Fɪɢ. 69.—*a*, larva of Ephemera with seven pairs of tracheal gills (*Kt*); *Tk*, an
isolated tracheal gill; *b*, tracheal system of an Agrion larva; *Tst*, tracheal
trunk; *Na*, accessory eyes (after Claus and Dufour).

oxygen carrier; the **heart** is a long tubular organ, lying in a
pericardium in the dorsal region, and it as a rule pumps the

blood into lacunar spaces. The blood may be colourless, yellow, red, or greenish.

The **Excretory system** consists of numerous thread-like processes from the hind-gut (see fig. 68, p. 107) which ramify amongst the organs, and which have been shown to contain uric acid Malpighian tubes.

Reproductive system.—The sexes are always distinct, and as a rule the females are longer, less active, and quieter in their colours than the males, from which they often further differ in not possessing wings. In the **male** there are paired **testes**, two ducts (**vasa deferentia**) which usually join together to form a single **ejaculatory** duct, one or two **seminal vesicles** for storing **spermatozoa, accessory glands**, and sometimes a copulatory organ (**penis**). In the **female** there are paired **ovaries**, two **oviducts**, generally a single **vagina** into which the oviducts open, a **receptaculum seminis** for storing spermatozoa, received from a male during copulation, and **accessory genital** glands.

Metamorphosis of insects.—Just as Crustaceans usually pass through one or more larval stages (Nauplius, Zooea, Mysis stages) before arriving at maturity, so also Insects generally have **larvæ** more or less dissimilar from the parent.

In many cases, however, the young only differ from their parents in not being sexually mature, in the absence of wings in winged forms, and in size. This is the case with Cockroaches, Earwigs, Lice, Green-fly (Aphides), Bugs, and some other Insects.

In Mayflies, Dragon-flies, etc., the larvæ are aquatic; they have closed tracheal apertures, and in most cases breathe by gills. In the Mayfly (Ephemeridæ) the larva has well-developed jaws, and feeds upon other Insects; it moults frequently, in some cases as many as twenty times during the larval stage, and during the last one or two moults, the larval structures become gradually absorbed and replaced by the structures special to the adult. Its aerial existence is very short, lasting only a few hours; whereas the larval stage extends over two years or more. The adult stage is

entirely devoted to the business of reproduction, the mouth parts being
only rudimentary and no nourishment being taken in. The Dragon-flies,
on the contrary, feed during both their larval and adult stages.

In the large majority of Insects (Flies, Fleas, Butterflies,
Moths, Beetles, Ants, Bees, etc.), there is a true metamorphosis;
the egg develops into a **larva** (maggot, grub, or caterpillar)
which does not bear any resemblance in appearance or habits
to the adult. The larval stage is essentially a feeding stage;
nourishment is taken up in large quantity, until the animal

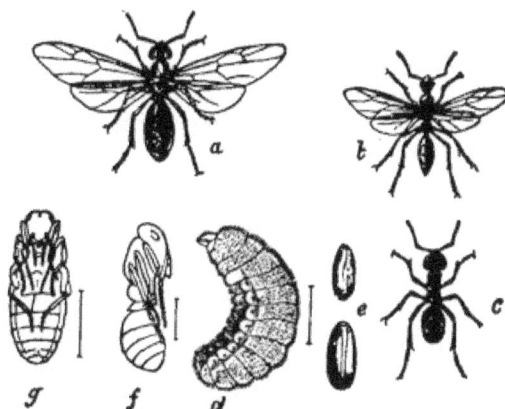

FIG. 70.—Formica herculanea (an Ant)(from Claus). *a*, female; *b*, male; *c*, worker; *d*, larva of Formica rufa; *e*, pupa with case; *f, g*, pupa liberated from the case.

becomes enormously fat; it then becomes quiescent, often
in a **cocoon**, as a **pupa, nymph, or chrysalis.** During this
apparently quiescent stage, the greater part of the body is
reconstructed; in winged forms wings bud out, adult appendages appear, colours are developed, the body alters its shape,
the internal organs become modified, then when all these
changes have taken place the adult animal emerges as an
imago. The male imago produces spermatozoa, the female
eggs, and so the life history is repeated. Most Insects

only have a short adult stage, and usually are very prolific ; if any species is to be perpetuated it is necessary that a very large number of eggs should be produced, for the complicated metamorphosis, during which reproduction does not of course occur, renders the animal peculiarly liable to have its life cut short. The necessity for storing up nourishment in readiness for the pupa stage, in itself makes the caterpillar, etc., peculiarly attractive to birds or other animals.

As far as possible to make up for this risk, many Insects, during either their larval or imago stages, take on various forms of protective adaptation, "walking leaves," "walking sticks," etc.; thus some that live on leaves are so similar in colour and appearance to them, that they may readily be overlooked. Some are well provided with weapons of defence, stings, biting organs, etc.; others, again, develop an unpleasant odour or taste, which makes them noxious to predatory animals.

A considerable number of Insects are parasitic, either as adults (Fleas, Lice, etc.) or as larvæ (maggots of Gadflies in cattle, boring larvæ in plants, etc.).

Many Insects do an enormous amount of damage, if their numbers are large, such as Locusts on crops, Phylloxera to the grape-vine, Green-fly to roses, etc. ; others again cause much irritation to their hosts (Fleas, Lice, etc.).

Arachnoidea.—This class includes Spiders, Scorpions, Mites, the King-crab (Limulus), and the Sea-spiders (Pycnogonidæ).

Arachnoidea, in most cases, can be at once distinguished from Insecta, in that, instead of there being three pairs of walking legs, there are **four pairs**. There are no appendages corresponding to the antennæ in Insects, nor as a rule have the abdominal segments any.

Scorpions have an unsegmented fused head and thorax (cephalo-thorax), an abdomen with twelve segments, and a well-developed **stinging organ**, containing venom, at the tip of the tail. The **respiratory organs** consist of chambered tracheæ (**lung books**). The **heart** lies dorsally, and consists of eight chambers.

The **Spiders** form an order containing a large number of members. Many form webs, and in that case are provided at the posterior part of the abdomen with glands (**spinning glands**), from the tubes of which a viscid material can be

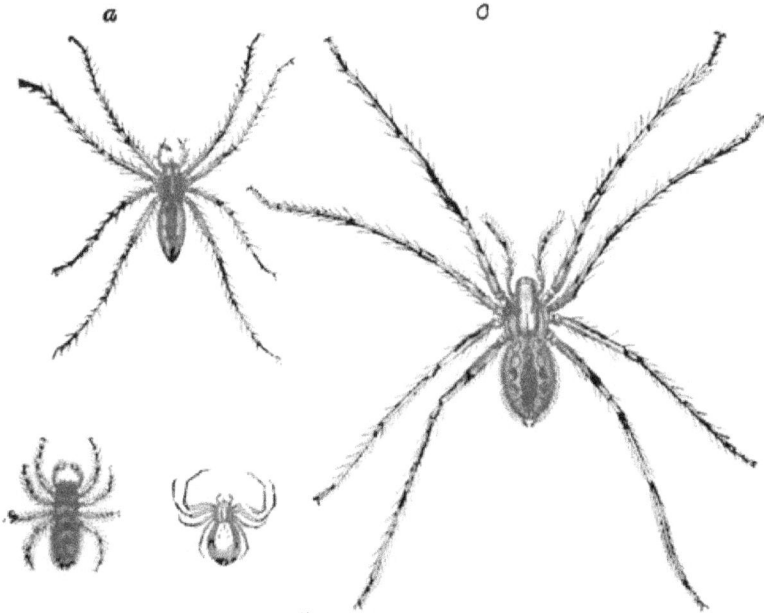

FIG. 71.—Female spiders (from Claus).

squeezed out. The animal, after pressing the openings of the tubes against the surface from which it is going to spin its web, moves away, with the result that the viscid material is drawn out into fine threads which are united together to form a single cord; the legs aid in extending and guiding the thread. The spider uses these threads, not only for forming a web, but also for lining its nest, and for forming a cocoon for its eggs.

8

The male spider is usually only about one-thirteenth the size of the female, and it often has considerable difficulty in depositing its spermatozoa in the vagina of the female, for if in any way it fails to please its mate, the latter attacks it, and, unless it is very agile in getting out of the way, kills it.

Spiders usually live upon insects which they catch in their webs and then suck "the juices" from out their bodies. The body consists of an **unsegmented cephalo-thorax**, an **unseg-**

Fig. 72.—Demodex folliculorum (after Mégnin). *Kt*, Pedipalp.

mented abdomen, and a narrow connecting part between them. Behind the gullet, there is generally a well-developed **suctorial** apparatus. The respiratory tracheæ are chiefly collected in chambers (the pulmonary sacs), but in addition in most forms there are tubular tracheæ.

Mites are very small Arachnoids, which are either parasitic or obtain their food from organic matters. The Itch-mite (Acarus scabei or Sarcoptes hominis), occurs as an external parasite on man. The Demodex folliculorum occurs very frequently in the sebaceous glands of man, especially in those in the skin of the nose. Gall-mites are not uncommon in plants. Harvest-mites also belong to this class.

Fig. 73.—Pentastomum denticulatum (from Claus). *O*, mouth; *Hf*, the four hooks; *D*, intestine; *A*, anus.

Pentastomata, which are also parasitic Arachnoids, are worm-like in form; their final host is usually the dog or the wolf, while their intermediate host is the rabbit or the hare.

The **King-crab** (Limulus) belongs to the Arachnoidea, differing very much in size, general appearance, and habitat from

other members of the class. It has a large cephalo-thorax, an
abdomen ending in a long spine, and five pairs of walking legs.
It breathes by means of plate-like gills borne upon appendages.

The **Mollusca** consist of three great classes—the **Lamellibranchiata** (Bivalved molluscs), the **Gasteropoda** (Snails), and
the **Cephalopoda** (Cuttle-fishes).

The majority of this class are marine animals, but some live
in fresh water and many on land. Whilst the Bivalves live
upon minute animals, the Snails are vegetarians, and the
Cuttle-fish are carnivors. Again, Bivalves are almost always
very sluggish in their movements, Snails are deliberate, and
Cuttle-fish are exceedingly active.

Lamellibranchiata (Bivalves).—If the ordinary fresh-water
Mussel (Anodonta Cygnea) be examined, it will be seen to be
enclosed in a shell with **two valves**. This shell is composed
chiefly of carbonate of lime, and is formed partly by the free
surface of the whole of the integument, but chiefly by the free
edges of the two **mantle lobes** which are beneath the shell,
and which envelope the rest of the body as with a mantle.
The main part of the body lies in the upper part of the
chamber formed by the two lobes of the mantle. Projecting from the body is a large muscular organ, **the foot**, by
means of which the animal can slowly crawl about. The
head region is very slightly developed, there being no
distinct part to which the name of head can fairly be given.
On the ventral surface of this region there is an opening, **the
mouth** ; this leads into the **stomach**. Opening out from the
stomach is the **intestine** ; this passes down into the foot ; it there
forms three loops, and then runs up again into the body ;
it then turns back, passes through a chamber representing the

body-cavity (the **pericardium**), and finally opens into a space at the posterior end of the ainmal (the **cloaca**). On either side of the stomach and opening into it are glandular organs, the **digestive glands**. **Respiration** is effected partly by means of the inner surface of the mantle, and partly by special plate-

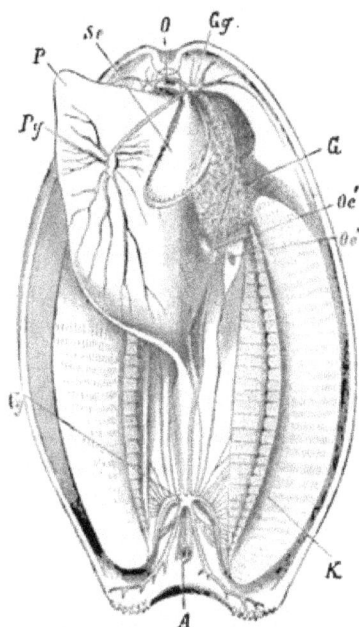

Fig. 74.—Swan mussel (*Anodon*) (after Kober). *O*, mouth; *A*, anus; *K*, gills; *P*, foot; *Se*, labial palps; *Gg*, cerebral ganglion; *Pg*, pedal ganglion; *Vg*, splanchnic ganglion; *G*, generative gland; *Oe'*, external opening of kidney; *Oe''*, opening of generative gland.

like organs, the **gills**; these hang down from the body between it and the foot, there being two on each side. Water passes in at the posterior end of the animal, below the junction of the two inner gills, by means of the inhalent siphon, carrying with it food substances. The ciliated cells on the surfaces of the gills and mantle lobes keep up a current of water, which passes around the mouth, and then up into spaces above the gills, and out again by an **exhalent aperture** at the posterior end above the gills.

The **heart** lies in the **pericardium,** and consists of a **ventricle** encircling the intestine (rectum), and **two auricles,** one on either side of the ventricle. The ventricle pumps blood into the different parts of the body, and this blood eventually returns from the mantle lobes and from the gills into the auricles and so to the ventricle.

The renal excretory organs (**organs of Bojamus**) are a pair of kidneys, each of which consists of a glandular part communicating with the pericardial cavity, and a muscular tube leading from the glandular part to the exterior.

The **Nervous system** consists of three ganglia (cerebral, pedal, and parieto splanchnic) and connectives between them. The muscle is all of the smooth variety.

The **Reproductive system** consists of a pair of testes or ovaries, lying in the upper part of the foot, and opening to the exterior by apertures close to the openings of the ureters.

The **Gasteropoda** include the air-breathing Snails and Slugs, and the aquatic Pond-snails, Periwinkles, Sea-hares (Aplysia), Limpets, etc. Some of these are symmetrical, in others the organs on the right side have, to a great extent, disappeared, those on the left side being well developed.

The animals in this class have usually a **single shell** ; the **head is well developed** ; the **mouth** contains a ribbon-like structure (**radula**), having a rough, rasping surface ; the **foot** is flattened except in the swimming forms, and the **digestive organs**, etc., are borne in a hump on the back.

The land Snails breathe by means of a **pulmonary chamber** well supplied with blood vessels. In some the sexes are separate ; others, like the Garden-snail, are hermaphrodite.

The **Cephalopoda** (Cuttle-fishes) are symmetrical, free swimming, marine animals. Some are enclosed in **shells** (Nautilus); others possess only a calcareous structure (the cuttle-bone of Sepia), a chitinous plate (Loligo), or nothing representing the shell of the ancestral mollusc (Octopus). They breathe by **gills**; their **nerve-ganglia** are enclosed in a cartilaginous box around the gullet; their **eyes** are generally well developed; the **vascular system** is well developed, and the sexes are separate.

CHAPTER XII.

VERTEBRATA.

THE **Vertebrata** are defined as animals which, at some stage of their existence, possess (1) a **notochord**, round which there is developed a sheath, which may or may not be replaced by a cartilaginous or bony skeleton; (2) **gill-slits**, which may or may not persist; (3) a **tubular dorsal nerve cord.**

Aristotle was the first to place most of the animals included under the name of Vertebrates together; he described them as a class consisting of animals with blood; he also considered the possession of a bony or cartilaginous vertebral column to be an important characteristic.

Linnæus gave, as the four highest classes of his system, beasts, birds, reptiles, and fishes.

Lamarck, in 1797, divided the whole animal world into two classes—animals with vertebræ, and animals without. Cuvier agreed with him in regarding the possession of vertebræ as distinguishing the Vertebrata from his three other great types.

Further research soon showed that the possession of bony vertebræ, that is, pieces of bone jointed to form a spinal column, could not be regarded as a distinguishing characteristic; for some fishes, which evidently belong to the class Vertebrata, do not possess a jointed bony vertebral column; thus the commoner Sharks and Skates possess purely cartilaginous skeletons, while Sturgeons and Lampreys have no vertebræ at all, but merely a continuous elastic rod in the place of the jointed spinal column. It was soon noticed that in these fishes the muscles and the connective tissue partitions were arranged as a series of segments attached to the sides of this elastic rod. It was then suggested that the presence of this segmental arrangemen of the muscles of the body-wall, as well as of a skeletal axis, which might be itself surrounded by an unsegmented connective tissue sheath, which again may be replaced by segmental cartilaginous or bony vertebræ, should be considered as the distinguishing characteristic of the Vertebrata.

The next important feature that was observed about Vertebrata was that they all possess at some period of their existence, laterally placed passages (gill-slits) leading from the pharynx to the exterior.

It was further shown that the great mass of nervous tissue lying dorsally above the spinal column, and known as the cerebro-spinal nervous system, or brain and spinal cord, is in all cases a tube, which originates as part of the dorsal surface of the embryo. This nerve tube arises in the form of a long groove, which becomes finally closed in by the adhesion of its opposite edges, a tube or canal thus being formed.

The definition of the Vertebrata which is now accepted includes animals which have only recently been shown to possess at some stage of their existence the three above-mentioned characteristic structures. These animals, in their adult condition, seem to have very little in common with the higher Vertebrata, but as they undoubtedly possess at one stage the vertebrate characteristics, they must be included under the general term. The Vertebrata are divided into four branches—

1. The Craniata (Cuvierian Vertebrates).
2. The Cephalochorda (Amphioxus).
3. The Urochorda (Tunicata).
4. The Hemichorda (Balanoglossus).

It is only of the Craniata that it is necessary to speak here.

The **Craniata** are Vertebrata, in which the tubular cerebro-spinal nerve mass is swollen anteriorly to form a brain. The notochord, whilst extending posteriorly to the extremity of the body, does not reach quite so far forward anteriorly as the termination of the nerve tube. Around the anterior extremity of the nerve cord, that is, round the brain, a cranium, or brain case, is developed for purposes of protection; hence the name Craniata.

The Skeleton.—The possession of an internal skeleton is of great importance. In the Invertebrata the firm supporting structures are almost always produced by the hardening and segmentation of the external skin. In the Vertebrata the relation of the hard to the soft parts of the body is reversed, the hard parts being placed in the interior of the body. These hard parts send out processes towards the dorsal and ventral surfaces, which constitute respectively a dorsal canal for the reception of the central nervous system, and a ventral arch over the vascular trunks and the viscera. As before mentioned, in some of the lower vertebrates, the axial skeleton in the adult is only an elastic rod, which is developed around the persistent notochord.

The notochord is present during embryonic life in the higher forms, but in the adult it becomes largely replaced by the bony vertebral column.   The bony vertebræ develop as rings in the sheath of the notochord.   These rings represent the first rudiments of the vertebral bodies, and in connection with them the arches are formed.

It must be carefully borne in mind that the notochord never develops into the backbone.   The spinal column—whether it consist of connective tissue as in the Amphioxus, of cartilage as in the Dog-fish, or of bone as in Man—is entirely formed from the mesoblastic sheath of the notochord, the notochord itself taking no share in its formation, but only forming a more or less persistent rod, around which it is developed.   Each vertebra consists of a body or centrum (in the centre of which remains of the notochord frequently persist), of a dorsal bony arch, enclosing a canal, in which the nerve cord lies, and of a ventral arch which, in some cases, encloses a canal in which blood vessels lie.   In addition to the above, transverse processes, which project from the sides of the bodies of the vertebræ, are developed.   The ribs are bony or cartilaginous rods, which are developed quite independently of the vertebral column, and which articulate by their dorsal ends, either with the processes which form the ventral arch as in Fishes, or as in the higher forms with the transverse processes, and the bodies of the vertebræ themselves.

In the anterior region of the body, as already mentioned, there are slits (the gill-slits) which lead from the pharynx to the exterior, and which persist throughout life in Fishes, but in air-breathing animals (such as man) only appear during embryonic life. They are absent in the adult, although traces of them can still be made out. Between the gill-slits, cartilaginous or bony rods develop in the body-wall for the purpose of supporting the vascular processes or gills, and of keeping the gill-slits open ; these rods develop on both sides of the animal, and their ventral ends unite together. They thus form a series of arches encircling the pharynx, and are known as the visceral arches, the whole arrangement constituting the visceral skeleton.

In the air-breathing forms, as there is no longer any need of bringing the blood vessels in the gills into contact with large quantities of water, so as to enable them to obtain sufficient oxygen for purposes of respiration, the gills disappear, the visceral clefts close up, and the visceral skeleton is almost entirely absent. In them it is only represented (1) by the lower jaw apparatus, which is usually armed with teeth ; and (2) by some of the small bones of the ear; and (3) by the hyoid bone with its anterior and posterior cornua.

In the lowest Vertebrates, the parts of the body behind the head may be divided into two regions. The anterior region or trunk contains the body-cavity, the various organs of digestion, the heart, the generative organs, etc. In the posterior region or tail, there is no body-cavity, and the ventral processes of the vertebræ (the hæmapophyses) unite together to enclose a canal which contains the main trunks of the tail blood vessels. These animals propel themselves by means of movements of their tails. In the higher forms, the animals move about by the aid of paired appendages, the fore and hind legs. In the forms between, such as most fishes, the paired appendages are modified to form fins which aid locomotion, at the same time the vertebral column remains freely movable.

In Snakes, which have no limbs, the ribs are freely articulated with the vertebral bodies, and can be moved backwards and forwards, while at the same time the vertebral column retains its mobility. They move partly by lateral flexions of the vertebral column, and partly by alternately back-ward and forward movements of the ribs, so that in a certain sense they may be said to run on the points of their ribs.

The **limbs** present wide differences in general appearance ;

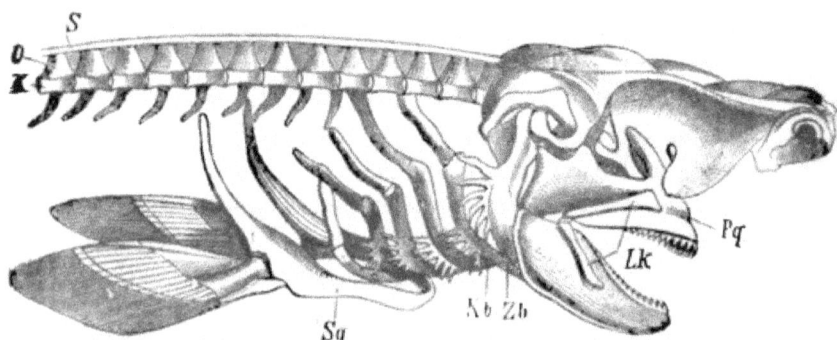

Fig. 75.—Head and anterior region of skeleton of Dogfish (after Owen). *K*, body of vertebra : *O*, neural arch ; *S*, intercalated piece ; *Pq*, palato quadrate ; *Lk*, labial cartilage ; *Zb*, hyoid arch ; *Kb*, branchial arch ; *Sg*, shoulder girdle.

they may have the form of legs, of wings, or of fins. In all cases, however, they consist of the same essential parts, the variation, suppression, or reduction of which determines the differences between them. Not only can the anterior appendages, whether they be wings, legs, or fins, be traced to a common type, but the hind and front limbs are essentially repetitions of the same arrangement.

There are two girdles, the Pectoral and Pelvic, with which the anterior and posterior appendages articulate.

Pectoral girdle.—The girdle with which the anterior pair of limbs articulate is called the pectoral girdle. It consists of three paired pieces ; two dorsal portions, the shoulder blades or **scapulæ**, and four ventral portions, the anterior or **præcoracoids** (in Man only represented by ligaments with which the clavicle becomes joined), and the posterior or **coracoids** (in man only represented by the coracoid processes). It is at the junction of the three portions of either side (the scapula, the præcoracoid, and the coracoid), that the **glenoid cavity** occurs for the articulation of the humerus with the girdle.

Pelvic girdle.—The posterior, or as it is called pelvic girdle, is usually fused more or less completely with the vertebral column by means of its dorsal portions. As the posterior limbs usually have to support the chief weight of the body, it is necessary that the connection between them and the vertebral column should be firm and strong ; this somewhat modifies the general appearance of the girdle. Still it is usually easy to make out on each side a dorsal portion, closely corresponding to the scapula of the pectoral girdle, called the ilium, and two ventral portions, corresponding to the præcoracoid and coracoid respectively, called the **pubis** and **ischium**. As in the pectoral girdle, it is at the junction of these three portions that the girdle articulates with its appendage on either side. The head of the femur fits into the **acetabulum**, into the formation of which all three bones enter.

In the cartilaginous Fishes, such as Sharks, the girdles are exceedingly simple, the pectoral and pelvic girdles each consist of two simple cartilaginous arches, which unite together in the middle line ventrally.

Anterior appendages.—The anterior limb in its simplest form consists of a jointed stem of cartilage, which articulates with the shoulder girdle. The cartilage is replaced by bone in all the higher forms.

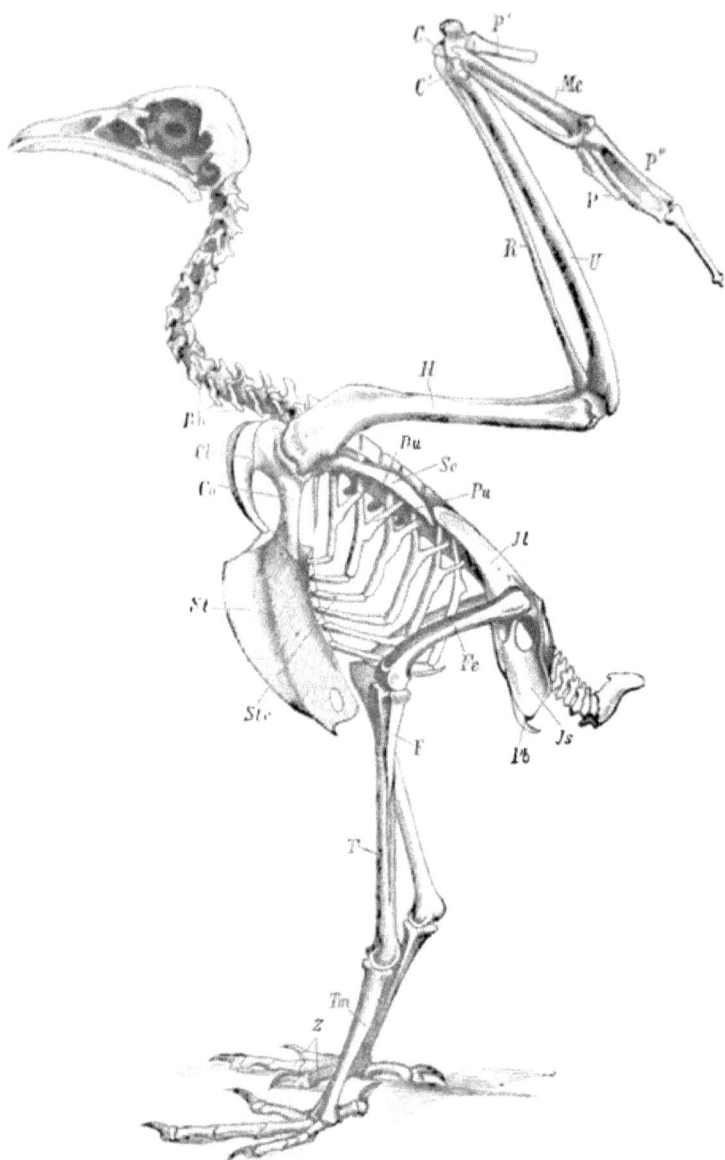

FIG. 76.—Skeleton of an Egyptian vulture (from Claus). *Rh*, cervical ribs; *Du*, inferior spinus process; *Cl*, clavicle; *Co*, coracoid; *Sc*, scapula; *St*, sternum; *Stc*, sternocostal bones; *Pu*, uncinate process of a thoracic rib; *Jl*, ilium; *Js*, ischium; *Pb*, pubis; *H*, humerus; *R*, radius; *U*, ulna; *C, C*, carpus; *Mc*, metacarpus; *P', P'', P'''*, phalanges of the three fingers; *Fe*, femur; *T*, tibia; *F*, fibula; *Tm*, tarso-metatarsus; *Z*, toes.

In Fishes, the jointed rod has attached to it, on either side, numerous smaller rods also jointed, the whole structure being covered with skin and constituting a fin.

In the higher Vertebrates, four segments, separated from each other by transverse joints, can be made out in the rod. These again may be further subdivided by longitudinal and transverse divisions. The first segment forms the **humerus** ; the second consists of the **radius** and **ulna**, which are sometimes, as in the Frog, fused to form a single bone ; the third includes the several bones of the **carpus** or wrist joint, and the fourth contains all the bones of the **manus** or hand. The hind limb corresponds closely with the fore limb, the **femur** corresponding to the humerus, the **tibia** and **fibula** to the radius and ulna, the **tarsus** or ankle joint to the carpus, and the **pes** or foot to the manus. The change in structure of the fin to a compound system of levers, such as is present in the higher Vertebrates, is due to a change in function, the swimming organ being modified to form a flying, running, or climbing organ.

In Birds, the fore limb is very similar to that of Man, as far as the humerus, radius, and ulna are concerned. The carpus is much reduced, the bones of the manus are much diminished in number, and are at the same time elongated and more or less fused together, so as to provide a suitable structure to support the wing feathers.* In flying Birds, the anterior limbs are always firmly attached to the shoulder girdle, which is well developed, and strengthened by the large clavicles.

In Animals that use their fore limbs for running, the carpus and manus present various minor modifications, the bones of the manus being more or less fused together.

Posterior appendages.—The hind limbs in the higher

---

* There are only two separate carpals, three metacarpals fused together, and three digits.

vertebrates present fewer modifications than the fore limbs do, as they more nearly subserve the same functions in different forms. In Birds, half of the tarsal bones are fused with the tibia to form the tibio-tarsus; the remaining tarsal and the metatarsal bones are also much fused together, and are generally much lengenthed, the bone thus formed is called the tarso-metatarsus, and it is due to the variations in length of this

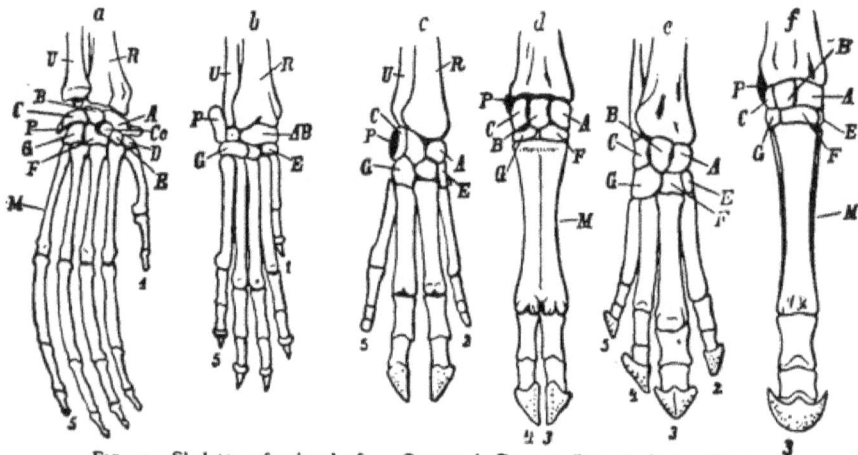

FIG. 77.—Skeleton of a hand of—*a*, Orang; *b*, Dog; *c*, Pig; *d*, Ox; *e*, Tapir; *f*, Horse (after Gegenbauer and Claus). *R*, radius; *U*, ulna; *A*, scaphoid; *B*, semi-lunar; *C*, cuneiform; *D*, trapezium; *E*, trapezoid; *F*, osmagnum; *G*, unciform; *P*, pisiform; *Cc*, centrale; *M*, metacarpus.

bone, that the great difference observed in the lengths of the legs of different birds is caused. The fibula is in Birds rudimentary and is fused with the tibia.

Ribs.—The ribs are bony arches developed in connection with the inferior processes of the vertebræ. They are usually articulated to the transverse processes of the vertebræ. These bones are sometimes only short or rudimentary; but usually they encircle the body, and enclose many of the internal organs. When well developed, they may become connected together on the ventral surface by means of the sternum. In some cases each rib consists of a bony rod which may be jointed or not; in other cases, as in Man, the ventral portion is cartilaginous, the dorsal portion being bony.

The **sternum** varies considerably in appearance in different Vertebrates; sometimes, as in Man, it is a comparatively simple plate of bone. In flying Birds, it is large and well developed with a projecting keel on the ventral surface which presents a large area for the attachment of the powerful muscles moving the wings.

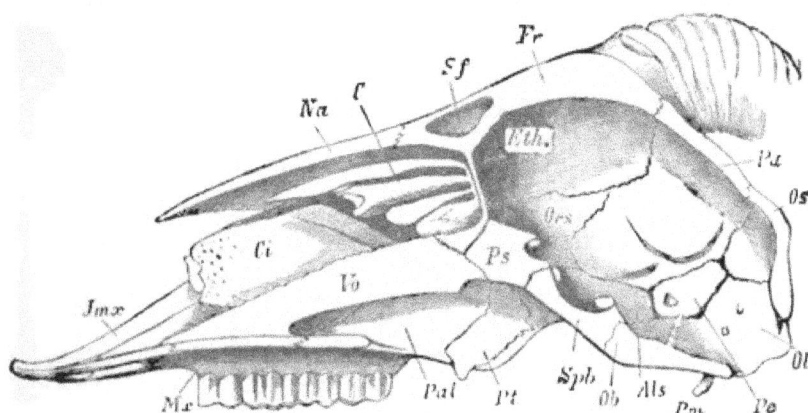

FIG. 78.—Median longitudinal section of a Sheep's skull (from Claus). *Ob*, basi-occipital; *Ol*, exoccipital; *Os*, supraoccipital; *Pe*, petrous bone; *Spb*, basi-sphenoid; *Ps*, præsphenoid; *Als*, alisphenoid; *Ors*, orbitosphenoid; *P* parietal; *Fr*, frontal; *Sf*, frontal sinus; *Na*, nasal; *C*, ethmo-turbinal; *Gi*, inferior turbinal; *Pt*, pterygoid; *Pal*, palatine; *Vo*, vomer; *Mx*, maxilla *Jmx*, præmaxilla.

Skull.—The skull, or brain box, is in the lower Fishes (such as Sharks) entirely composed of cartilage, but in all the higher Vertebrates the cartilage is more or less entirely replaced or covered in by bones.

In the embryo it consists of four basal plates of cartilage—a posterior pair (the parachordals), with which the anterior extremity of the notochord is generally fused, and an anterior pair (the trabeculæ cranii).   From these basal plates, processes of cartilage grow up on either side to partly enclose the brain.   The box is completed anteriorly by the cartilaginous framework of the organs of smell (the olfactory capsules).   Posteriorly the processes meet on the dorsal surface of the hinder part of the cranium, to form the occipital series of bones which partly roof in the cavity posteriorly.   Just in front of the occipital segments, the framework enclosing the auditory organs (auditory capsule on either side) helps to complete the box.   The roof of the box is completed by bones called membrane bones, which are not first laid down in cartilage, but only in membrane.   In all the higher Vertebrata the cartilage becomes replaced by bone, and membrane bones are developed whenever a gap is left between the cartilage bones.   The two visceral arches, which lie farthest forward, also assist in building up the skull, the mandibular arch forming the basis of the lower jaw and of the pterygoid and palatine bones of the upper jaw ; and the upper part of both the hyoid and the mandibular arches being concerned in the formation of some of the bones of the ear.   The lower part of the hyoid arch forms the main part of the hyoid bone.

To summarise:—From the parachordals and the trabeculæ cranii, the occipital, sphenoid and ethmoid bones are formed ; the nasal sense capsules help to close in the skull anteriorly and laterally, whilst the auditory sense capsules assist in closing it in posteriorly and laterally ; the mandibular and hyoid arches form the lower jaw, part of the upper jaw, the hyoid bone, and part of the ear apparatus ; and the membrane bones fill up all the gaps left, the nasals, frontals, parietals forming the greater part of the roof, the lachrymals, squamosals, etc., forming part of the sides, the præmaxillæ, maxillæ, etc., forming the upper jaw, and several membrane bones helping to complete the formation of the lower jaw.

Integument.—The skin in Vertebrates is divided into two distinct layers, the **epidermis**, externally, which is composed of cells flattened on the surface and spherical in the deeper layers, and the **true skin or corium**, which is composed of fibrous connective tissue.   The skin, muscles are in connection with the corium.   In man muscles are only slightly developed in connection with the skin, but in many vertebrates they are of considerable importance, serving to move the skin and its appendages. Hairs, nails, and feathers are developed entirely from the epidermis ; scales and teeth are partly derived from the epidermis and partly from the deeper structures.

The Digestive Apparatus.—The **alimentary canal** is chiefly derived from the hypoblast, but at either end there is an inpushing of epiblast, forming anteriorly the mouth (**stomatodœum**), and posteriorly the anus (**proctodœum**). The **teeth** are partly derived from epiblast, but chiefly from mesoblast, the dentine or ivory being derived from the latter, whilst the cap of enamel is derived from the former.

Teeth are not always present ; thus, in adult Birds, in Tortoises, and in a few low Mammals, these organs are replaced by a horny substance which covers the sharp edges of the jaws (beak). In some Whales the teeth, which have commenced to form in the embryo, become absorbed before birth, and horny plates (the so-called whalebone) develop as processes from the palate. In the poisonous Snakes there are special teeth in the upper jaw traversed by a groove or canal, through which the poison from the poison glands passes into the wounds.

The **tongue,** and in the higher forms the **salivary glands,** are both associated in development with the mouth.

The tongue is richly supplied with nerves, and functions largely as an organ of taste ; but in some animals, as in the Frog, it serves also as an organ for catching food, and in all animals it is of importance for the taking in of food. The saliva secreted by the salivary glands has a double function ; it converts non-diffusible starch into diffusible sugar, by means of a digestive ferment which it contains, and it moistens the food. The importance of this moistening action may be gathered from the negative observation of the absence of salivary glands in Fishes.

The mouth leads into the **pharynx,** which is perforated in Fishes by the gill-slits, so that water taken in by the mouth can pass out by the gill-slits. In the higher forms, in which the

gill-slits have disappeared, the pharynx only has opening into
it, the mouth cavity, the posterior openings of the nasal cavity,

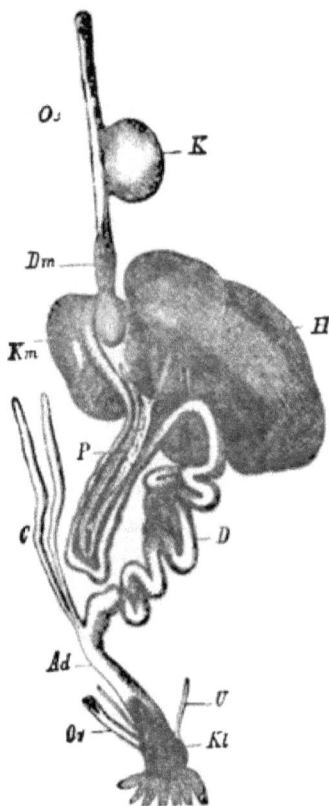

Fig. 79.—Digestive canal of a Bird (from Claus). *Ov*, œsophagus; *K*, crop;
Dm, proventriculus; *Km*, gizzard; *D*, small intestine; *P*, pancreas; *H*, liver;
C, the two cæca; *Ad*, large intestine; *U*, ureter; *Ov*, oviduct; *Kl*, cloaca.

and the Eustachian tubes, and opening from it are the entrances
to the œsophagus or gullet and to the larynx. Associated
with the pharynx in the region of the gill-slits are two organs,
whose functions are very imperfectly understood, namely, the

thyroid body and the thymus gland. In the adult the latter atrophies in the higher Vertebrates. The **œsophagus** or gullet, which varies in length according to the length of the neck, leads direct into the stomach. It is, as a rule, a simple tube, but in many Birds there is a dilatation upon it, the **crop**, in which the food is softened. The lower end of the œsophagus in these animals is dilated into a glandular organ, proventriculus, which is functionally part of the stomach, as in it the gastric juice is secreted.

The **stomach** is a digestive organ; it has numerous simple glands in its wall, which pour their secretion into its cavity. In Birds and in Crocodiles it is an exceedingly muscular organ, the muscular walls being especially thick in grain-eating Birds, and in these it is called the gizzard.

The **intestine** is distinguished by its great length and by the presence of numerous folds and projections of the membrane lining it, the function of which is to increase the surface over which digestion and absorption can take place. This surface is still further increased in some animals (as Birds, Herbivora, etc.) by the existence of one or two tubes of varying length which open into the alimentary canal at the junction of the small and large intestine. These tubes are closed at their free ends and are called cæca. The length of the small intestine varies, being very long in vegetable feeders, shorter in pure animal feeders, and of medium length in omniverous animals, like Man.

This variation in length is well illustrated by the Frog, the Tadpole being a vegetable feeder, while the adult Frog lives upon animal food, such as insects. In the Tadpole the intestine is exceedingly long, being coiled up like a watch-spring; but as the animal grows and changes its habits, the intestine does not grow at the same rate, the result being that the flesh-eating Frog possesses a relatively short intestine.

Opening into the small intestine are the ducts of the **liver** and the **pancreas**; the former is a large organ which regulates the composition of the blood and which pours bile into the intestine; the latter secretes a digestive juice. The small intestine opens into the large intestine, which in its turn leads into the rectum, which opens to the exterior by the anus.*

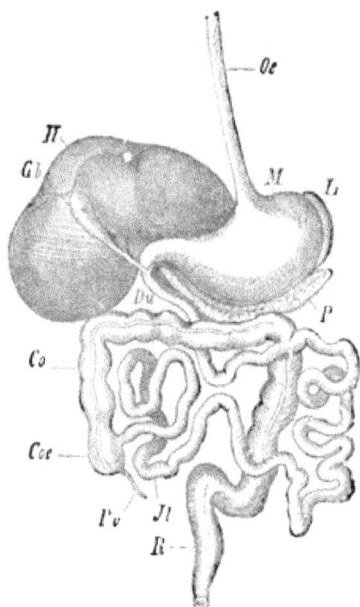

FIG. 80.—Alimentary canal of Man (from Claus). *Oe*, œsophagus; *M*, stomach; *L*, spleen; *H*, liver; *Gb*, gall bladder; *P*, pancreas; *Du*, duodenum receiving bile and pancreatic ducts; *Jl*, ileum; *Co*, colon; *Coe*, cæcum with vermiform appendix (*Pv*); *R*, rectum.

In most of the higher Vertebrates the fæces (indigestible substances, and used-up secretions, etc.) are discharged

---

* The lower part of the alimentary canal is in some animals divided into two parts which are called the large **intestine** and the **rectum**; in others no such division can be made out, the whole being spoken of indifferently by either name.

directly to the exterior at the anus. In some Fishes, Frogs, Birds, and Snakes, the **rectum** opens into a cloaca or common sewer, which receives not only the fæces but also the urinary excretion and genital products. The cloaca opens to the exterior at the anus. The position of the anus varies with the length of the tail; it is always on the ventral surface, and is usually at the hinder part of the animal. In a few Fishes it opens well forward at the junction of the anterior third with the posterior two-thirds of the body.

Organs of Respiration.—Special respiratory organs, either **gills** or **lungs**, are present in nearly all vertebrates. It is only in Fishes and Amphibians that gills are found. The gills in the lowest Fishes consist of much-folded membrane abundantly supplied with blood vessels. In the higher Fishes they consist of branchial filaments which are borne upon the visceral arches, and project into the visceral clefts (gill-slits).* Water passes freely in through the mouth and out by the gill-slits. The blood is aerated by the oxygen contained in the water. In many Fishes, an organ (the swimming bladder) arises as an outgrowth from the alimentary canal. This is generally an unpaired sac which sometimes retains its connection with the alimentary canal by means of an air tube, the pneumatic duct, but sometimes is only a closed sac. It is supposed to help the fish in altering its specific gravity, but as some fish swim very well without it, it is evidently not necessary to enable the fish to rise or sink in the water. In a small class of Fishes, the Dipnoi, double breathers or mud Fishes, the swim bladder is copiously supplied with blood vessels, and functions as a lung, so that these animals have a double system of respiratory organs.

---

* The gills are often protected by a fold of skin, or by a more elaborate operculum.

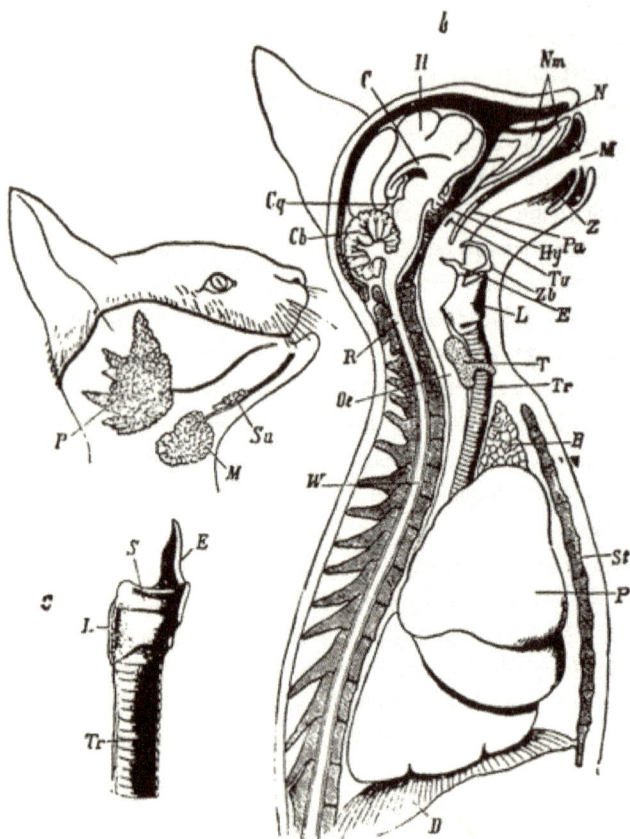

FIG. 81.—Entrance to the digestive and respiratory organs of the Cat (after Heider). *a*, head; *P*, parotid; *M*, submaxillary and *Su*, sublingual salivary glands; *b*, *N*, nasal aperture; *Nm*, turbinal bones; *M*, mouth; *Z*, tongue; *Pa*, velum palati; *Oe*, œsophagus; *L*, larynx; *E*, epiglottis; *Zb*, hyoid; *Tr*, trachea; *P*, lung; *D*, diaphragm; *T*, thyroid; *B*, thymus; *Tu*, opening of Eustachian tube into pharynx; *H*, cerebral hemispheres; *C*, corpus callosum; *Cq*, corpora quadragemina; *Cb*, cerebellum; *R*, spinal cord; *Hy*, hypophysis; *W*, vertebral column; *St*, sternum; *C*, larynx (*L*) and first part of trachea (*Tr*); *S*, vocal cord; *E*, epiglottis (longitudinal section).

It appears at first sight that the swimming bladder of the fish is only a modified lung, but this is probably not the case.

Lungs occur in Reptiles, Birds, Mammals, and adult Amphibians. They develop as an outgrowth of the alimentary canal. This outgrowth becomes completely separated off from the œsophagus, and at its lower end divides into two or more tubes, which communicate with the pharynx by a single tube, the **trachea**. The lungs, in their simplest form, consist of dilated sacs which are richly supplied with blood vessels, but in the higher Vertebrates they form spongy bodies, consisting of large numbers of minute air sacs or vesicles, with blood vessels ramifying on their walls. These air vesicles open by small tubes or bronchioles into larger tubes or bronchi, which in turn open into the trachea. In Snakes, the lung is nearly always much reduced in size on one side, that on the other side being correspondingly enlarged. In Birds, there are large diverticula, or air sacs, which communicate with large air spaces in the bones and between the muscles, so that the animal contains within the body a large amount of air, and thus is rendered as light as possible. In the higher Vertebrates (as Man) the lungs are freely suspended in the thorax, that is in the anterior portion of the body-cavity, this being shut off from the posterior part by a sheet of muscular and tendinous tissue, the **diaphragm**. At the upper end of the trachea or windpipe, just before it communicates with the pharynx, there is a vocal apparatus or **larynx**. This vocal apparatus consists of two folds of membrane, the edges of which can be tightened, approximated to or separated from each other; these are the **vocal cords**. This apparatus is well protected and rendered rigid by a cartilaginous framework. In some animals the upper opening of the larynx (**the glottis**) can be more or less completely closed by a flaplike structure, the **epiglottis**.

In most Birds, there is a more or less complicated vocal apparatus placed at the lower end of the trachea, just where it divides into the two main bronchi (the bifurcation of the trachea). This vocal apparatus in Birds is called the lower larynx or syrinx.

Circulatory system.—The blood of all Vertebrates is a fluid, in which white amœboid corpuscles float. In addition to the amœboid corpuscles the fluid (except in the lowest

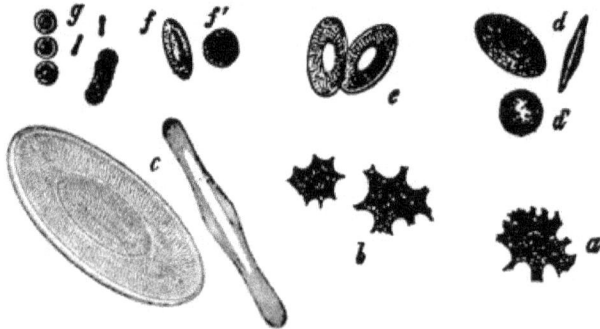

FIG. 82.—Blood corpuscles (after Ecker). *a*, colourless corpuscles from the Swan mussel; *b*, from a Caterpillar; *c*, red corpuscles from Proteus; *d*, from the Smooth adder; *d'*, lymph corpuscles of the same; *e*, red corpuscles of the Frog; *f*, of the Pigeon; *f'*, lymph corpuscles of the same; *g*, red blood corpuscles of Man.

Vertebrates) contains an enormous number of red corpuscles (five millions to the cubic millimetre in Man). The red corpuscles are usually oval nucleated discs, except in Mammals, where, with the exception of the Camel tribe, they are circular and contain no nuclei. They owe their colour to the hæmoglobin they contain; this substance has a powerful affinity for oxygen which it takes up from water or air as the case may be, and carries to the tissues, and there gives it up to the protoplasm of the cells.

The vascular system consists essentially of a closed system

of tubes, in one part of which is inserted a muscular organ,
the heart, which lies in a special part of the body-cavity, the

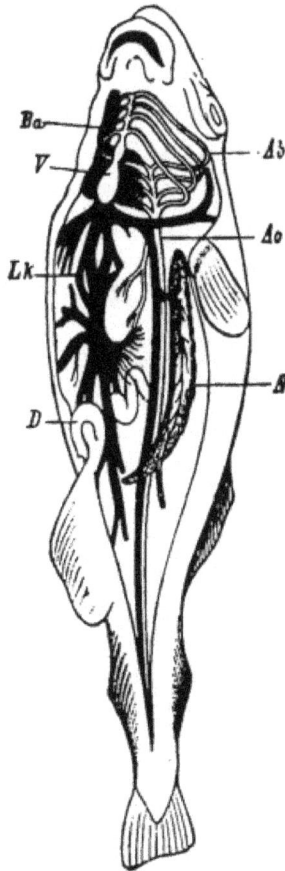

F IG. 83.—Diagram of the circulatory organs of an osseous Fish. *V*, ventricle;
   Ba, aortic bulb with the arterial arches which carry blood to the gills;
   Ao, dorsal aorta formed by the vessels from the gills (*Ab*); *N*, kidney;
   D, alimentary canal; *Lk,* portal circulation (from Claus).

pericardium. All vessels returning blood to the heart ar e
described as veins, whilst those bearing it away from the heart

are called arteries ; these names are used quite irrespective of whether the blood is aerated and so bright red, or reduced and so dark red.

In all Fishes, except those possessing both lungs and gills (the Dipnoi or double breathers), the heart possesses only two chambers—an **auricle** into which the blood from the veins is poured, and a **ventricle** which drives it on by alternate contractions and relaxations into the main artery.

In most cases the large veins just at their junction with the auricle dilate to form a third chamber called the sinus venosus ; a fourth chamber, the truncus arteriosus, is similarly formed by the main artery as it leaves the ventricle.

The main artery soon divides into a number of branches, the different gill vessels, which go to the gills ; these branches break up into a large number of minute tubes, the capillaries, which ramify in the gill plates, and thus bring the blood into close relationship with the oxygen in the water. The capillaries then collect into a small number of larger vessels, the efferent vessels of the gills, which, after sending off branches to the head, join to form a large vessel, the dorsal aorta. This vessel runs along the dorsal surface of the body, giving off branches as it proceeds to the fins, the body-walls, the alimentary canal, etc., until it ends as the caudal artery in the tail. Each arterial branch given off from this vessel breaks up into capillaries in the organ to which it goes, thus carrying aerated blood to every part of the organ. The capillaries re-collect to form veins, and these carry the non-aerated blood to the heart, pouring it into the auricle. In all Vertebrates the course of the venous blood from the greater part of the alimentary canal to the heart is interrupted in the liver. The blood, laden with the products of digestion, is brought to this organ by the portal vein which breaks up into capillaries so as to bring the blood into close relationship with the liver cells. The capillaries then re-collect to form large vessels, the hepatic veins, which carry the blood on to the heart. This is described as the hepatic portal circulation.

In Fishes, Amphibia, most Reptiles, and to a less extent in Crocodiles, Tortoises, and Birds, most of the blood returning from the posterior parts of the body passes through the kidneys, a renal portal circulation being thus formed. The blood, as it passes through the kidneys, gives up its nitrogenous waste

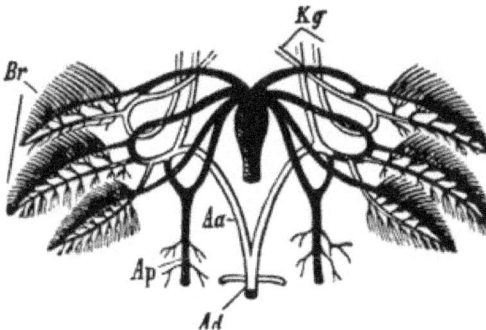

FIG. 84.—Aortic arches of Tadpole (after Bergmann and Leuckart). *Aa*, aortic arches uniting into descending aorta (*Ad*); *Ap*, pulmonary artery; *Kg*, cephalic arteries; *Br*, gills.

material, and this is excreted by the kidneys. It then passes by the renal veins into the system of large veins which carry the blood back to the heart.

In animals which, like the Frog, breathe at one time by gills, and at another by lungs, the course of the circulation in the young gill-breathing stage, is similar to that above described, as occurring in most Fishes ; and whilst the animal breathes by gills only, as in the young Tadpole, this arrangement is the only one that exists ; but later, when the animal begins to breathe by lungs instead of gills, it becomes somewhat modified ;

the anterior gill arteries instead of going to the gills pass directly into the systemic arteries, whilst the posterior gill arteries instead of going to the gills pass direct to the lungs, and from the lungs the aerated blood is returned direct to the heart. At the same time the auricle becomes divided by a septum into two chambers—one of these, now called the left auricle, receives only aerated blood from the lungs ; whilst the other, the right auricle, continues to receive the impure blood from the rest of the system. The ventricle still remains single ; it hence receives both pure (aerated) and impure (non-aerated) blood, and a more or less complicated arrangement is developed, so as to send the greater part of the impure blood to the lungs (and in some cases, as in the Frog, also to the skin) to be aerated, whilst the pure blood is sent to the head ; the pure and impure blood, which have unavoidably become mixed, going to the rest of the system.

In most Reptiles, the ventricle is single, but it is partially divided by a septum into two chambers, so that only a slight admixture of the pure and impure blood takes place.

In Birds and Mammals, the ventricle is completely divided so that the heart consists of four chambers, a right auricle and ventricle, and a left auricle and ventricle. The impure non-aerated blood all flows into the right auricle, thence to the right ventricle, from there it passes to the lungs where it is aerated ; it is then returned to the left auricle, from which it flows into the left ventricle. From the left ventricle it passes by means of the dorsal aorta to the whole of the system, being returned after it has given up its oxygen, and got rid of its nitrogenous waste products in the kidneys, etc., direct to the right auricle.

This double system is described as the **lesser or pulmonary** circulation and **greater or systemic** circulation. The former consists of the pulmonary arteries to the lungs from the right ventricle, pulmonary capillaries in the lungs, and pulmonary

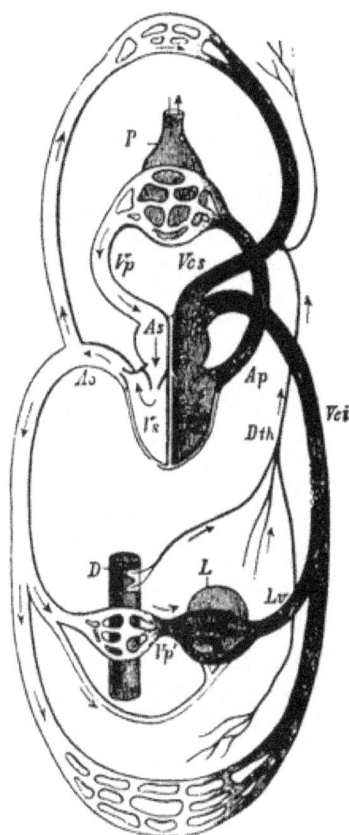

Fig. 85.—Diagram of the circulation in an Animal with a double circulation (after Huxley). *Ad*, right auricle; *As*, left auricle; *Vd*, right and *Vs*, left ventricle; *Vcs*, superior and *Vci*, inferior vena cava; *Dth*, thoracic duct; *Ap*, pulmonary artery; *Vp*, pulmonary vein; *P*, lung; *Ao*, aorta; *D*, intestine; *L*, liver; *Vp'*, portal vein; *Lv*, hepatic vein.

veins to the left auricle. The latter consists of the dorsal aorta from the left ventricle with its branches and their capillaries, and the hepatic veins and superior and inferior venæ cavæ, all of which open into the right auricle.

Lymphatic system.—**Lymph** is a colourless fluid in which the colourless lymph corpuscles float. **The lymphatic system** is a subdivision of the general circulatory system, and is essentially a drainage system. It must not be understood, however, by this term, that the fluids conveyed by it are of no further use to the organism. Every part of the body is supplied with a system of vessels (lymphatics) which commence as spaces between the cells (the lymph spaces). Into these spaces flow the nutritive fluids, which have been poured out by the blood system upon the cells. In the walls of the alimentary canal, these vessels take up some of the digestive products, especially fats ; the fluid they contain then presents a milky appearance, and is called chyle, the vessels themselves being described as **lacteals**. The lymph spaces open into minute vessels which collect into larger trunks and eventually empty themselves into the veins. The main trunk of the lymphatic system runs along the ventral surface of the vertebral column, and is called the thoracic duct. In the higher Vertebrates, this duct opens into one of the veins of the neck, but in the lower there are numerous communications between the vascular and lymphatic systems. In their course the lymphatics are interrupted by glands which act as filters to strain off any hurtful substances. These substances either remain stored up in the glands, or more generally are destroyed by the activity of the amœboid cells in the glands.

Urinary organs.—Except in the lowest Vertebrates, the urinary organs consist of a pair of **kidneys**, which are generally placed in the posterior part of the abdominal cavity. Each kidney may be considered to consist of a collection of nephridia or segmental organs (similar to those seen in the earth-worm).

The kidneys pour their urinary secretion into a pair of ducts or **ureters**, which run down, one from each kidney to open into the bladder, into a sinus common to the genital and the urinary products, or into the cloaca.

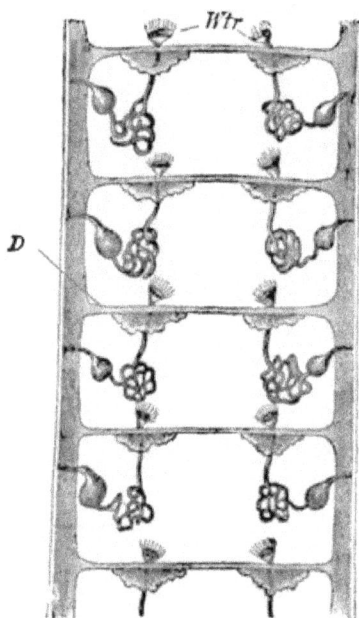

FIG. 86.—Diagrammatic representation of the segmental organs (*nephridia*) in a segmented Worm (after C. Semper). *Ds*, septum; *Wtr*, ciliated funnels of nephridia.

The ureter develops on the surface of the embryo as a groove which later becomes formed into a canal, so that although the nephridia or tubules do not open to the exterior, they open into a canal which in development is pinched off from the exterior, its walls being epiblastic in origin.

In connection with each tubule or nephridium, a pouch appears, into which a coil of arterial blood vessels (glomerulus) becomes pushed, the pouch with its blood vessels being known

10

as the Malpighian body.   As a rule the ciliated funnel, which
was present during development, becomes completely closed.
The walls of the tubule for a considerable part of their extent
are lined with glandular cells, and are richly supplied with
blood vessels.   The urine is derived partly from the blood as

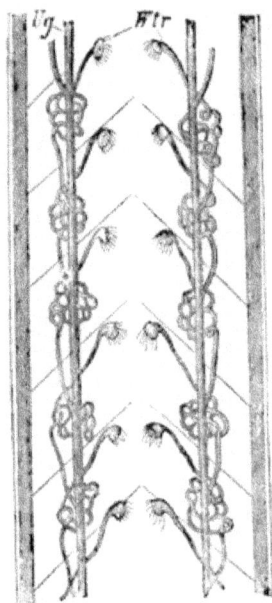

FIG. 87.—Diagrammatic representation of the kidney (segmental organs) of a Dog-
fish embryo (after C. Semper).   *Wtr*, ciliated funnels; *Ug*, kidney duct.

it passes through the glomeruli and partly from that in the
vessels between the tubules.   The glandular cells, which line
the walls of these tubules, take up from the blood the nitrogenous waste material, and excrete it into the duct of the
tubule ; it is then washed down to the exterior by the fluid
which has been poured out by the vessels in the glomerulus.

The Nervous system.—This consists of a **brain,** of a **spinal**

cord which contains a narrow central canal, and of the cranial and spinal nerves. The central canal of the spinal cord is continued into the brain where it is dilated to form the ventricles. The brain is only the dilated anterior end of the spinal cord; it consists primarily of three dilatations, called the three **primary vesicles**, and it is from these that all the parts of the complex adult brain are developed. The **first primary vesicle** gives rise anteriorly to the **olfactory lobes**, and laterally to the **cerebral hemispheres.** The remainder of this vesicle is called

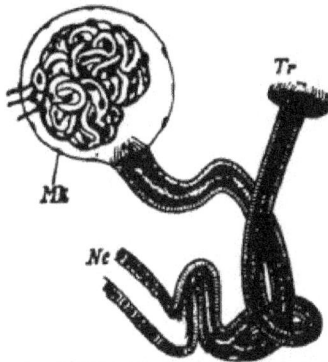

FIG. 88.—Ciliated funnel and Malpighian body from the anterior part of the kidney of Proteus (after Spengel). *Nc*, kidney tubule; *Tr*, ciliated funnel; *Mk*, Malpighian body.

the **thalamencephalon**; on the dorsal surface of this there is a small outgrowth called the **pineal gland**, supposed to be the rudimentary remains of a median eye, and on its ventral surface there is another outgrowth, the **pituitary body.** The dilatation of the anterior end of the spinal canal in the thalamencephalon is called the **third ventricle**, and this communicates with the two **lateral ventricles** in the cerebral hemispheres. Posteriorly the third ventricle communicates with the canal connecting the third with the **fourth ventricle.**

The **second primary vesicle** develops less than does either

the first or third.   In the lower Vertebrates it becomes divided
into two **optic lobes**, each of which contains a small ventricle,
while in the higher Vertebrates each optic lobe becomes
partially divided into two, the four lobes thus produced being
called the **corpora quadrigemina**.

The **third primary vesicle** gives rise to an outgrowth, which
in some animals (as the Frog) is small and unimportant, whilst
in most animals it is large and well developed.   It forms the

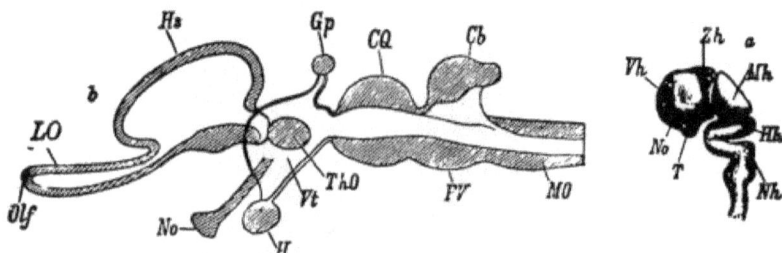

Fig. 89.—*a*, brain and anterior part of spinal cord of human embryo (after
   Kölliker): *Vh*, fore brain; *Zh*, thalamencephalon; *Mh*, mid brain; *Hh*, hind
   brain; *Nh*, medulla oblongata; *No*, optic nerve; *T*, anterior ventral ending
   of the thalamencephalon. *b*, diagrammatic longitudinal section through a
   vertebrate brain (after Huxley): *Hs*, cerebral hemispheres; *LO*, olfactory
   lobes; *Olf*, olfactory nerve; *ThO*, optic thalamus; *Vt*, third ventricle; *No*,
   optic nerve; *H*, pituitary body; *Gp*, pineal gland; *CQ*, corpora quadrigemina;
   Cb, cerebellum; *MO*, medulla oblongata; *PV*, pons varolii.

cerebellum and **pons varolii.**   The remainder of this vesicle
forms the **medulla oblongata** which is continuous posteriorly
with the spinal cord.   The medulla oblongata contains within
it the fourth ventricle.

The various parts of the brain present great variations in
size and complexity, according to whether the animal is low or
high in the scale of development ; thus, in the lower Fishes the
cerebral hemispheres are small and are only incompletely
separated from each other, whereas in the higher Vertebrates
they are so large that they overlap and conceal beneath them

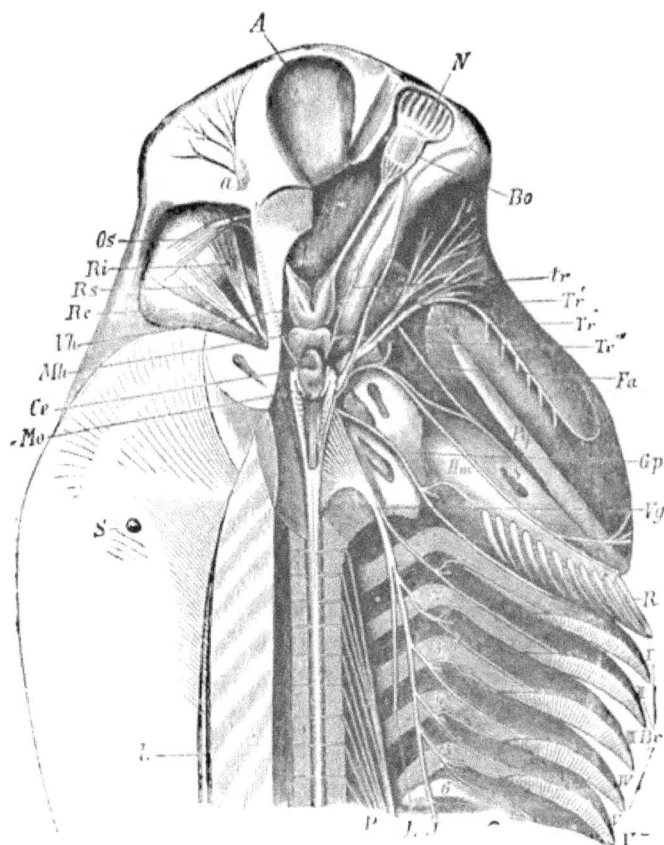

FIG. 90.—Brain and anterior part of spinal cord of Hexanchus griseus (after
Gegenbauer). Right eye has been removed. *A*, anterior cavity of the skull;
N, nasal capsule; *Vh*, cerebral hemispheres; *Mh*, optic lobes; *Ce*, cerebellum;
Mo, medulla oblongata; *Bo*, olfactory bulb; *tr*, fourth nerve; *Tr'*, ophthalmic
branch of fifth nerve; *a*, terminal branches of same; *Tr''*, second and *Tr'''*,
third branch of fifth; *Fa*, seventh nerve; *Gp*, ninth; *Vg*, tenth; *L*, branch of
tenth to lateral line; *J*, intestinal branch; *Os*, superior oblique muscle of eye;
Ri, internal, *Re*, external, and *Rs*, superior rectus muscles; *S*, spiracle;
Pq, palato quadrate; *Hm*, hyomandibular; *R*, branchial rays; *I—VI*, branchial
arches; *Br*, branchiæ; *P*, spinal nerves.

all the other parts of the brain. The thalamencephalon, the optic lobes, the cerebellum, and the medulla oblongata vary less than do the cerebral hemispheres. The higher the animal, the more complex does the brain become, reaching its most complex condition in Man.

The **cranial and spinal nerves** arise as outgrowths from the central nervous system. In the nerves arising from the spinal cord, two roots—a single dorsal **sensory root**, and a single or multiple ventral **motor root**—occur. In the brain many of the nerves primitively arise by a single dorsal root. The first ten cranial nerves are very constant in their distribution. The first, the olfactory, supplies the nasal organ ; the second, the optic, is the nerve of sight ; the third or oculo motor, the fourth or pathetic, and the sixth or abducens, are all motor nerves for the muscles moving the eyeball ; the fifth or trigeminus sends one large branch forwards through the orbit, a second to the upper jaw, and a third to the lower jaw; the seventh or facial chiefly supplies muscles, and in Fishes forks over the hyoidean gill-slit ; the eighth or auditory goes to the ear ; the ninth or glossopharyngeal supplies the pharynx, and in Fishes forks over the first gill-slit ; and the tenth or vagus goes to the gills or lungs, to the heart, and to the stomach.

The spinal nerves arise as outgrowths from the spinal cord; the dorsal or posterior are always single as in the cranial nerves, and arise from a continuous ridge or crest along the dorsal surface of the cord ; each root has generally a swelling upon it containing nerve cells and called the ganglion of the posterior root. The anterior or ventral roots are frequently multiple at their origin and never present any ganglionic swellings upon them. It has, however, been suggested that the ganglia upon the visceral or sympathetic nervous system, which is formed of nerve branches derived from the anterior roots, represent the ganglia of these roots.

The sympathetic nerves are derived from the anterior roots of the spinal nerves ; they supply the viscera and the blood vessels.

Sense organs.—The sense organs develop, like the central

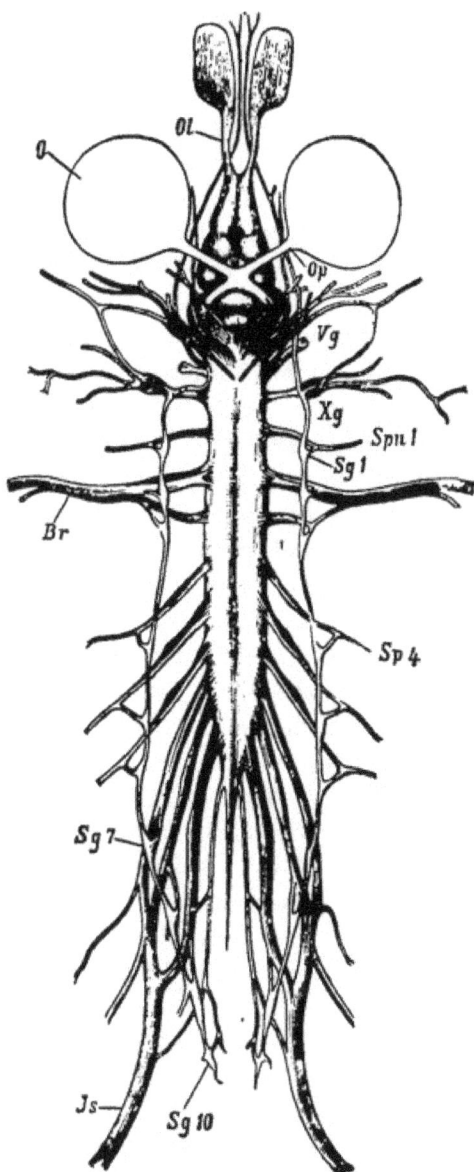

FIG. 91.—Nervous system of the Frog (after Ecker). *Ol*, olfactory nerves; *O*, eye; *Op*, optic nerve; *Vg*, Gasserian ganglion; *Xg*, ganglion of vagus; *Spn* 1, first spinal nerve; *Br*, branchial nerve; *Sg* 1—10, the ten ganglia of the sympathetic system; *Js*, ischial nerve.

nervous system, from the external layer of the embryo, that is from the epiblast. The ingrowths from the epiblast come into relationship with the sensory nerves or outgrowths from the brain, and thus sensations received by the sense organs from the external world are transmitted to the brain. The **nasal sense organs** develop as pits, and in connection with the sensory epithelium of these pits, the endings of the olfactory nerves are found. **The ear** develops in a similar manner, but in the higher Vertebrates the pit sinks deeply into the auditory capsule, and its narrowed neck closes. It also becomes subdivided into a complicated series of chambers and canals into connection with which come the branches of the auditory nerve. In most Fishes there is no direct path for impressions of sound to travel to the auditory organs, but in all the higher Vertebrates the spiracular or hyomandibular gill-cleft becomes replaced by an outgrowth which forms a cavity. This, the tympanic cavity, is shut off from the exterior by the tympanum or drum of the ear. The internal part of the cleft is represented by an outgrowth from the pharynx to the internal ear (the Eustachian tube).

The organ of vision, **the eye**, is present in almost all Vertebrates; in a few of the lower forms it is absent or degenerate, and in some that always live in darkness, it is hidden under the skin. In all the higher forms it is well developed, and consists of an ingrowth from the epidermis which forms the lens, and an outgrowth from the brain, the retina, which is connected with the brain by the optic nerve. In addition there are numerous blood vessels and various layers of pigmented epithelium, muscle, etc., all of which are derived from the middle layer or mesoblast.

Reproductive organs.—With the exception of a few Fishes the sexes are separate, the male being provided with a pair of

testes, and the female with either one or, in most cases, a pair of **ovaries.** In some low Fishes, the male and female genital products (**spermatozoa** and **ova**), after being thrown off from the testes and ovaries into the abdominal cavity, pass to the exterior through a genital pore placed behind the anus; but in most Vertebrates they pass to the exterior by special ducts. In

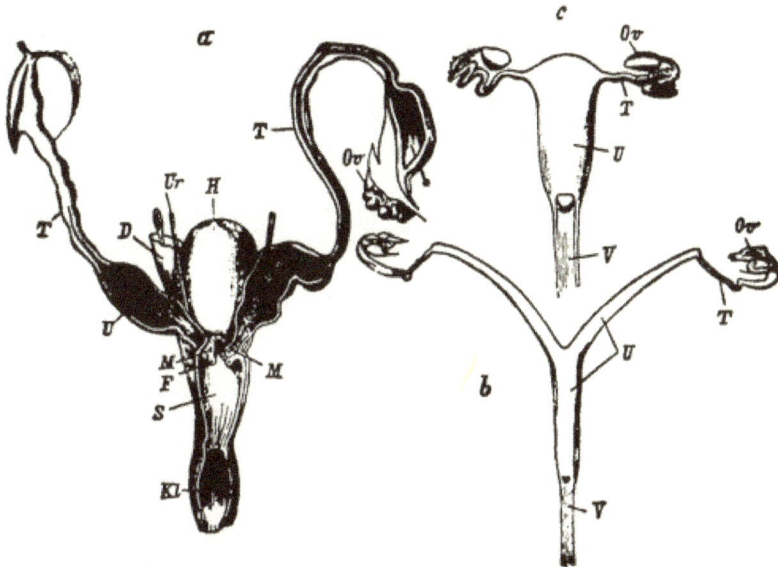

Fig 92.—Female generative organs. *a*, Ornithorhynchus (after Owen); *b*, of a Civet; *c*, of an Ape (from Claus). *Ov*, ovary; *T*, oviduct; *U*, uterus; *V*, vagina; *H*, urinary bladder; *Ur*, ureter; *M*, mouth of uterus; *F*, opening of ureter; *S*, urogenital sinus; *Kl*, cloaca; *D*, intestine.

Amphibia (Frogs, etc.) the spermatozoa pass down the ureters; but in most other adult Vertebrates the genital and urinary ducts are separate as far as the bladder, the urino-genital sinus, or the cloaca. In some Fishes the ovary is single, and in Birds the right ovary and oviduct are either much reduced in size or are entirely absent. In most Mammals the testes move from their original abdominal position into a pouch of skin (the scrotum).

The female genital duct (the oviduct or Müllerian duct) has been generally

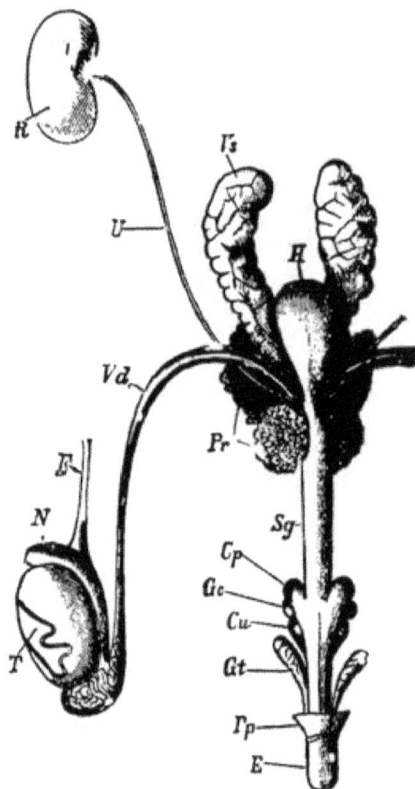

FIG. 93.—Urinary and sexual organs of a Mouse (*Cricetus vulgaris*) (after Gegenbauer). *R*, kidney; *U*, ureter; *H*, urinary bladder; *T*, testis; *F*, spermatic
cord; *N*, epididymis; *Vd*, vas deferens; *Vs*, vesicula seminalis; *Pr*, prostate;
Sg, urethra; *Gc*, Cowper's glands; *Gt*, Tyson's glands; *Cp*, corpora cavernosa;
E, glans penis; *Pp*, prepuce.

considered to be the persistent duct of a temporary head-kidney (pronephros);
but the view has recently been brought forward that this duct disappears
in the female as well as in the male, and that the oviduct is developed from
a fold of the peritoneum.

The male genital duct is the persistent duct of the middle kidney (mesonephros, Wolffian body), which in the Amphibia forms the ureter, but in other Vertebrates functions only as the male efferent duct, the Wolffian body itself functioning as epididymis.

The eggs formed in the ovary are set free into the body-cavity by the rupture of the membrane covering the ovary; they then pass directly into the openings of the oviducts. The lower portions of the two oviducts frequently join together to form a single dilated tube, which, in the higher Vertebrates, is known as the uterus. The eggs in almost all mammalia lodge in the uterus, and there undergo development, the young being born alive.

In many Fishes and Amphibia, the unfertilised eggs from the female and the spermatozoa from the male are both cast into the water, fertilisation of the eggs thus taking place outside the animal.

In most other Vertebrates, there are more or less complicated copulatory organs, the male being provided with a **penis**, along which the duct from the bladder (the urethra) passes. The spermatozoa are discharged into this duct, and during copulation are directly introduced into a pouch, the **vagina**, with which the **uterus** communicates.

CHAPTER XIII.

AMŒBA.

THE Amœba is a **Protozoon**, that is to say, it is an animal consisting of a single cell. Like all the Protozoa, it is very small.

The term Protozoon, when first introduced about forty-three years ago, was somewhat vaguely used. Various organisms, now excluded, such as some multicellular animals and many unicellular plants, were all included under the generic name of Protozoa. Now, however, the term is restricted to unicellular animals, all those consisting of more than a single cell being classed as Metazoa.

It is true that some Protozoa form colonies, that is to say, a number of them cohere together, and thus appear to form a multicellular animal; but in all such cases the colony can be broken up, and each detached cell will then lead an independent, unimpaired existence, fulfilling all its own functions, and living on exactly as though it were still attached to its fellows. This shows that the colony really consisted of a number of independent unicellular animals simply cohering together.

The Amœba is one of the simplest of the Protozoa. It consists of a mass of naked protoplasm, which is capable of taking solid particles of food into its substance and digesting them.

Amœbæ usually live on the surface of the mud at the bottom of fresh-water pools. The largest are of sufficient size to be just visible to the naked eye, whilst the smallest are less than $\frac{1}{300}$ inch in diameter.

One of the most characteristic points about the Amœba is

its ever-changing shape; and it was owing to this, that when
first discovered, the name proteus animalcule was given to it.

Anatomy.—The protoplasm of the Amœba is viscid, and
not very much denser than the water in which it is living. On
careful examination it can be readily seen that the protoplasm

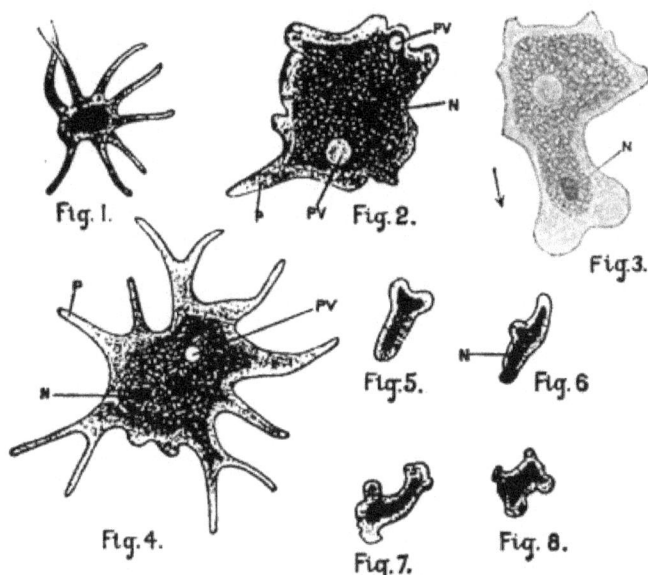

FIG. 94.—Amœba (from Marshall and Hurst). Figs. 2 and 3 were drawn from the
same specimen; figs. 5, 6, 7, and 8 were drawn from one specimen at intervals
of about twenty seconds. *N*, nucleus; *P*, pseudopodium; *PV*, pulsating
vacuole.

can be divided into two parts—an inner more granular portion,
and an outer clearer portion. The former is described as the
endoplasm or **endosaro**, the latter as the **eotoplasm** or **eotosaro**.

As before mentioned, the Amœba is constantly changing its
shape, and this change of shape may be seen to be accompanied
by the protrusion and retraction of blunt processes, called
pseudopodia. It is only the clear part of the protoplasm

whioh flows into these protrusions, no granules being visible in
them.   The pseudopodia may be protruded from any part of
the surface, and by means of them a slow, crawling movement
is produced.   The animal throws out a pseudopodium, fixes
the blunt end to the surface on which it is crawling, and then
draws its body up to it.

The endosarc contains several clear spaces or **vacuoles**,
which are filled with a clear fluid, less dense than the surrounding protoplasm.   These vacuoles frequently contain
small masses of food, and are then called **food vacuoles**.

Pulsating, or **Contractile vacuole.**—At one part of the
animal a vacuole may be seen which is fairly constant in its
position, and which starts as a minute speck.   It gradually
grows larger and larger until it appears as a clear space of considerable size.   It then suddenly disappears.   Its gradual
increase in size is due to its taking up water into itself from the
surrounding protoplasm.   This water has been absorbed through
the ectoplasm, either with food or without.   Its sudden disappearance is due to the bursting of this water through the
ectosarc, and the consequent flowing in of the surrounding
protoplasm to take its place.

The larger granules in the endosarc may consist of one of
three substances—of food, of substances from which all the
nourishment has been extracted, or of substances produced by
the degeneration of the protoplasm.

The **Nucleus** is not very easy to see unless it be stained.   It
is a spherical or ovoid body consisting of protoplasm, which is
denser than the endosarc.   As a rule there is only a single
nucleus, but sometimes two or even three have been observed.

Physiology.—The Amœba feeds upon minute plants or
animals which it takes into its body in a peculiarly characteristic manner.   As it has no mouth, food particles can only get

in by being forced through the ectosarc. The animal throws out one or two pseudopodia which encircle the food, together with a drop of water. The surface of the ectosarc in contact with the food then gives way, so that the food, still encircled by its drop of water, is precipitated into the protoplasm. A food vacuole is thus formed; the water soon disperses, leaving the particle of food in direct contact with the protoplasm. The water from the food vacuoles, together with that absorbed by the protoplasm through the ectosarc, collects into one spot to form the pulsating vacuole which, as before mentioned, appears, enlarges, and disappears with some regularity. The pulsations of the contractile vacuole indicate a more or less regular washing out of the protoplasm, thus ridding it of soluble waste matters such as are excreted in the higher animals by the kidneys. The contractile vacuole of the Amœba, therefore, may be said to be an **excretory organ.**

The food particles, having come into direct contact with the protoplasm, undergo digestion. All the substances that they contain, which are capable of building up protoplasm, are extracted and assimilated, and the remainder are thrown out from the animal through the ectosarc in a manner similar to that by which the food was taken in. Exactly how digestion goes on is not very clear, but, as the food is directly taken into the animal, the process is described as one of **intracellular digestion.**

Although the Amœba possesses no trace of a nervous system, yet it has some power of discrimination as to its food. This is shown by the fact that an Amœba will not take into its substance everything with which it may come into contact. Those substances which would be hurtful to it, it rejects. In addition to solid food particles and water, the protoplasm of the Amœba, like all other protoplasm, requires oxygen for the continuation of its activity; this it obtains from the water in which it lives, and which contains oxygen in solution.

The solid food particles consist usually of minute animals or plants.   We thus see that the Amœba feeds upon protoplasm, and thus it obtains its carbon, hydrogen, and nitrogen in a highly elaborated form.   In that, it differs widely from the unicellular chlorophyll-containing plants.   These plants, however, resemble the Amœba in obtaining their oxygen from the water in which they live.

If the surface on which the Amœba is lying be roughly shaken, or if any body in the water comes into contact with the animal, all the pseudopodia are withdrawn, and the Amœba takes on a spherical shape, and remains quiescent for some time until the disturbing influence is withdrawn.

Reproduction.—The Amœba reproduces itself by **fission.** When the animal has grown so large as to become unwieldy, it divides into two parts.   It is probable that the nucleus is the first to divide.   Thus two Amœbæ are produced each exactly resembling the original one, each consisting of ectosarc, endosarc, nucleus, etc.

This simple mode of reproduction is the only one that has been definitely proved to occur constantly in Amœbæ. Occasionally, however, two Amœbæ fuse together, surround themselves with a dense membrane, and then break up into numerous little pieces, each of which eventually grows up into an adult animal. This mode of reproduction is an instance of conjugation.

CHAPTER XIV.

YEAST PLANT (TORULA, OR SACCHAROMYCES CEREVISIÆ).

THE **Yeast Plant** is a minute unicellular Fungus. It grows either on the surface of saccharine fluids or immersed in them.

Anatomy.—A typical Yeast plant consists of a cell provided with a delicate **cellulose cell-wall.** It is round or oval in shape, and from $\frac{1}{7000}$ to $\frac{1}{3500}$ inch in diameter.

FIG. 95.—Growing cells of yeast (*Saccharomyces cerevisiæ*) (after Prantl).

The **cell-wall** encloses the **protoplasm**, imbedded in which there is a body presenting the characters of a **nucleus.** This may be demonstrated by careful staining.

One or more small **vacuoles** containing cell-sap usually occur in the protoplasm. The protoplasm contains no starch, but a small quantity of fat is usually present. The Yeast plants either occur singly or collected together in heaps or chains.

Reproduction.—The Yeast plant multiplies in two ways :—

1. By **gemmation** or **budding.**
2. By **endogenous spore formation**.

Budding.—One or more minute buds are thrown out, which increase in size, become separated from the parent by the formation of the cell-wall, and eventually become detached. A considerable number of plants may cohere together for a considerable time forming the above-mentioned masses or chains.

Endogenous Spore Formation.—If Yeast be cultivated upon a porous plate, or on the cut surface of a potato, some cells instead of throwing out buds will be seen to form spores. The protoplasm usually becomes divided into four spores, each of which surrounds itself with a cell-wall. Eventually these spores are set free by the rupture and disintegration of the cell-wall of the parent. The latter mode of reproduction only occurs if the conditions of growth are rather unfavourable, the usual method of reproduction being by gemmation.

Under favourable conditions yeast will multiply with such extreme rapidity, that if a few plants be added to a clear fluid which contains suitable nourishment, and if this clear fluid be kept for a few hours in a warm place, such an enormous number of yeast plants will develop in it, that the previously clear fluid will become turbid.

Physiology.—As Yeast contains no chlorophyll, it cannot obtain its carbon from the carbon dioxide of the atmosphere ; hence all its processes go on as well in darkness as in light ; its carbon is obtained from organic substances, and its nitrogen in a manner similar to that of other plants, namely from comparatively simple nitrogenous salts present in solution in the water around it. If free oxygen be present, the yeast absorbs it in a manner common to both plants and animals, and gives off carbon dioxide. If no free oxygen be present the Yeast plant seems to be able to obtain sufficient to maintain life from complex oxygen-containing bodies, but in the absence of free oxygen both growth and multiplication are far less rapid.

One of the most characteristic properties of the Yeast plant is its power of setting up alcoholic **fermentation** in fluids containing sugar. The fermentation processes go on much more rapidly when free oxygen is absent than when it is present. The sugar ($C_6 H_{12} O_6$) becomes split up, with the result that alcohol ($C_2 H_5 OH$) and carbon dioxide (CO_2) are produced.

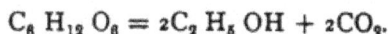

$$C_6 H_{12} O_6 = 2C_2 H_5 OH + 2CO_2.$$

Most of the sugar is thus split up, but a small amount is probably assimilated by the plant itself, and about 5 per cent. breaks up into glycerine and succinic acid.

Various other Fungi, and certain parts of the higher plants, such as ripe fruits, may be induced to set up an alcoholic fermentation in sugar in the absence of free oxygen.

For Yeast to grow and multiply, to maintain the activity of its protoplasm, to form new protoplasm, cellulose, cell-wall, fat, etc., it must be supplied with all the necessary builders up of protoplasm ; that is to say, with **Carbon, Oxygen, Hydrogen, Nitrogen, Sulphur, Phosphorus, Potassium, Magnesium,** and **Calcium.** As it contains no chlorophyll it must like other Fungi have its carbon supplied to it in an organic form.

CHAPTER XV.

PROTOCOCCUS AND GLEOCAPSA.

THE **Protococcus viridis** forms a good example of the unicellular Algæ, and in common with other Algæ contains chlorophyll. It lives in moist situations, such as in gutters, on the trunks of trees, or on damp roofs.

An adult specimen of Protococcus is a spherical body, varying in size, but always very minute. It is provided with a **cellulose cell-wall** enclosing a mass of **protoplasm**, in which is imbedded a **nucleus**, containing a **nucleolus**. In the protoplasm, in addition to the nucleus, there can readily be seen to be a number of small bodies, coloured bright green by the chlorophyll they contain ; these are called **chromatophores**, and correspond to the chlorophyll corpuscles of the higher plants. The chromatophores are so arranged that they form a lining on the inner surface of the cell-wall. The minute plants are either isolated or collected together in groups. **Reproduction** is very simple. A single plant becomes divided into two by means of a partition wall ; the nucleus at the same time divides, so that two new plants are produced, each one consisting of cell-wall, protoplasm, chromatophores, and nucleus.

Another form of Protococcus, namely, the **Protococcus or Hæmatococcus pluvialis**, is not quite so common, but it is very frequently found in water-butts, shallow pools, etc. ; it varies in size from $\frac{1}{250}$ to $\frac{1}{4000}$ inch in diameter. At one stage, it is very like the Protococcus viridis, but it differs from it, in that its cell-wall is thicker, and that the chlorophyll within the chromatophores is frequently coloured red instead of green. Small proteid bodies (pyrenoids), surrounded by a layer of starch, frequently occur in the chromatophores. Before arriving at the resting stage, the Protococcus pluvialis passes through a motile stage, in which its cellulose cell-wall is thin, and is separated from the protoplasm by a space containing

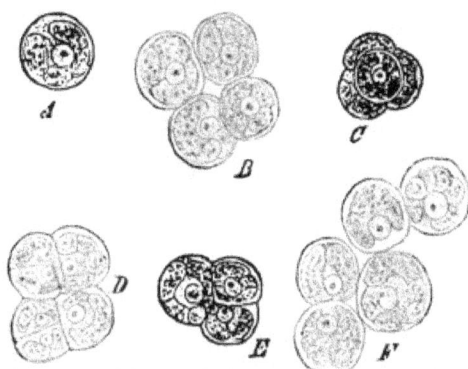

FIG. 96.—Protococcus viridis (from Strasburger). *A*, single plant showing chromatophores and nucleus; *B*, *C*, *D*, *E*, and *F*, several plants united together; in *D* the cells on the left have just divided.

cell-sap. The plant is provided with two long vibratile flagella, which are prolongations of the protoplasm at the pointed end of its pear-shaped body, and which traverse the space between the protoplasm and the cell-wall, perforating the latter. By means of the rapid movements of these flagella, the plant can move actively about, travelling at about the rate of a foot in a quarter of an hour. Before reproduction occurs, the resting phase is always assumed, the cell-wall becomes thickened, and, lying closely in contact with the protoplasm, forms a cyst around it.

The **reproduction** of Protococcus pluvialis differs considerably from that described as occurring in Protococcus viridis, it multiplies by endogenous spore formation; two kinds of spores are produced, which differ from each other in size, the larger being called **macrospores** and the smaller **microspores**. As both kinds are motile, they are both described as **zoospores**. In a plant which is going to form **macrospores**, the protoplasm divides itself first into two and then into four portions, each of which assumes a pear-like shape. The macrospores thus formed are then set free by the absorption of the cell-wall of the parent. Each consists of a naked mass of protoplasm, provided at its small end with two vibratile flagella. After a short time the protoplasm develops around itself a cellulose cell-wall, and thus the plant assumes the form of an adult Protococcus pluvialis in the motile stage. After a time the plant comes to rest, develops a thicker cell-wall, loses its flagella, and thus forms an adult resting Protococcus pluvialis.

In a plant which is going to form **microspores** the protoplasm becomes divided into eight portions, the cell-wall of the parent is absorbed, the microspores are set free and develop flagella just like the macrospores, but during the whole of their free swimming stage they remain naked, only developing a cell-wall when they come to rest.

Physiology.—The protoplasm in the Protococcus is more highly differentiated than that in yeast, hence it may be said to have reached a higher stage of development. Differentiation and consequent division of labour have taken place, part of the protoplasm forming the chromatophores, part giving rise to the cell-wall, and part being formed into flagella, which enable the plant during one stage of its existence to move actively about.

In its mode of nutrition the Protococcus resembles other green plants. It lives entirely on oxygen, on carbon dioxide, and on simple salts which are held in solution in the water in which it is growing. The carbon dioxide is decomposed, the carbon being made to unite with water, with the result that starch $C_6H_{10}O_5$ is formed and oxygen set free—

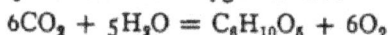

$$6CO_2 + 5H_2O = C_6H_{10}O_5 + 6O_2$$

The starch can be frequently seen in the protoplasm surrounding the pyrenoids. This process can only go on in the presence of sunlight. That oxygen is set free can be demonstrated by exposing tne plants in water to sunlight, collecting some of the bubbles of gas that are given off from them, and testing the gas so obtained ; it will be found to be oxygen. In common with all other protoplasm, whether animal or vegetable, that of Protococcus needs oxygen to enable it to perform its functions. Oxygen is being constantly taken up by the plant, and a corresponding quantity of carbon dioxide is being constantly given off, that is to say, a process of respiration is continually going on. During the day the amount of oxygen given off is far in excess of that required for the processes of respiration, and correspondingly, much more carbon dioxide is decomposed by the protoplasm than is formed by it. Respiration goes on however in the dark, whilst no carbon dioxide is decomposed, and no oxygen is given off.

The salts necessary for the growth of the plant must contain sulphur, phosphorus, potassium, magnesium, and calcium, as with yeast, but in addition the Protococcus, in common with all chlorophyll-containing plants, requires iron for the purpose of forming the green or red colouring matter of the chlorophyll. The zoospores are sensitive to light, swimming toward the light if it be not too strong, and away from it if it be very intense.

Gleocapsa.—Growing on flower-pots and on the glass of greenhouses there is frequently found another unicellular Alga, namely, Gleocapsa.

Anatomy.—The Gleocapsa is a **unicellular** Alga belonging to the family of the Chroöcoccaceæ. It possesses a **striated, gelatinous cell-wall**; it contains **chlorophyll** which is diffused through the protoplasm, and so far no nucleus has been discovered in it.

Reproduction.—The globular cell-body of the Gleocapsa lengthens until it becomes oval, the protoplasm then becomes

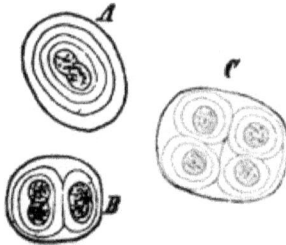

Fig. 97.—Gleocapsa (from Strasburger). *A*, at the commencement of division; *B*, shortly after division; *C*, four cells enclosed within a single outer cell-wall.

constricted in the middle, and soon divides into two. A delicate cell-wall then grows around each of the two masses of protoplasm. The portion of the cell-wall in contact with the protoplasm thickens and becomes rounded off, so that two plants are enclosed within the original cell-wall. These daughter cells again divide with the result that four or eight plants may all be enclosed by a common wall, which is the remains of the original cell-wall. A considerable number of generations are therefore combined together to form a cell colony. After a time the external wall ruptures and the separate plants are set free; almost immediately division again occurs in each of these isolated plants, so that it is rare to find a single Gleocapsa, two or more being almost always found united together.

Physiology.—In its mode of nutrition, the Gleocapsa resembles the Protococcus, obtaining its carbon from carbon dioxide in the presence of sunlight, its oxygen for respiration from the atmosphere, and its other constituents from simple salts in solution in the moisture in which it lives.

At no stage of its existence does the Gleocapsa ever form motile zoospores.

CHAPTER XVI.

BACTERIA.

BACTERIA are unicellular Thallophytes. They are generally considered to belong to the fission Fungi, or **Schizomycetes**, but as three forms are known to contain chlorophyll, it is perhaps better to consider them as a class by themselves.

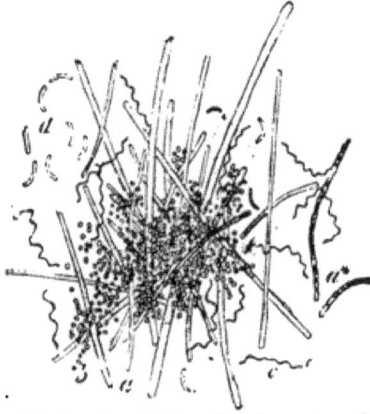

FIG. 98.—Bacteria of the fur of teeth (from Strasburger). *a*, Leptothrix buccalis in *a*ʺ after treatment with iodine; *b*, Micrococcus; *c*, Spirillum dentium, after treatment with iodine; *d*, "Comma bacillus" of the mucous membrane of the mouth.

Bacteria are exceedingly minute, unicellular, spherical or thread-like **plants**, which multiply by fission. The large majority of them contain no chlorophyll, and hence must obtain their carbon from more or less complex organic carbon-containing substances. The few chlorophyll-containing forms obtain it from carbon dioxide. All the other constituents of their food are obtained in the form of solutions of simple salts. They have a remarkable power of breaking down complex chemical substances into simple ones; they, like the Yeast plant, sometimes accomplish this by setting up **fermentation** processes in them. These organic substances are split up so that the resulting simple ones may serve the Bacteria for food.

This power that Bacteria possess, of breaking up complex organic substances, is of the utmost importance in the economy of nature. Were it not for them, the surface of the earth would soon be covered with the bodies of its dead inhabitants. When an animal or plant dies, all the complex organic substances, of which it is composed, must be broken up into simple salts before they can be made use of by ordinary plants This breaking up is chiefly performed by animals, but it is also largely brought about by the agency of Bacteria. We thus see, that Bacteria are most powerful auxiliaries to animals.

The fermentation action is set up by substances produced by the activity of the protoplasm of the Bacteria. The substances formed vary in different species, and the effects produced are correspondingly different.

Amongst the results produced by Bacteria by processes of fermentation, the following may be mentioned : starch is converted into sugar, cellulose is dissolved, albumen is peptonised, being converted from an insoluble to a soluble form, urea is converted into carbonate of ammonium, lactic acid is formed in milk, etc.

Putrefaction, or as it has been called **putrefactive fermentation**, is a complicated process, varying considerably according to the amount of oxygen present. It is usually a double process, the first stage of which is brought about by the Bacteria which need abundant oxygen (aërobic species), and the second by those which flourish when oxygen is absent (anaërobic species). If oxygen be present in abundance, the putrefactive process is not as a rule characterised by the evolution of foul-smelling gases ; if oxygen be absent or limited in amount, true putrefaction occurs, and various foul-smelling gases, together with various substances (ptomaines) which are mostly of a highly poisonous nature, are produced. The final result of the processes set up is, that the complex organic substances are decomposed into simple substances.

Many Bacteria produce vivid colours during their growth.

The Bacteria which lead a **parasitic** existence very frequently set up diseases in their hosts ; thus tuberculosis, erysipelas, anthrax, and a large number of other diseases in man are due to the presence of these micro-organisms. A Bacterium consists of protoplasm enclosed by a **cell-wall,** in which colouring matters are occasionally imbedded. In most forms the protoplasm is colourless, and fat globules are frequently present in it. Bodies resembling nuclei have been lately seen in some forms, but their exact nature has not been finally determined.

Outside the cell-wall a **gelatinous sheath** occurs, which is developed to a very variable degree in different forms.

Some Bacteria are always at rest, the trembling movements frequently seen in them, being only such as are exhibited by non-living particles. They are described as Brownian movements, and are due to mechanical causes. Others are sometimes at rest and sometimes in active movement. In some cases the movements are rotatory; in others again they are more definite, and are set up by flagella, which project from one or both ends of the cell (some species of Spirilla and Bacilli). The movements are sometimes slow, oscillating and rolling, in others they are rapid and darting.

In the quiescent state the Bacteria are either isolated, or collected together in long threads or chains; occasionally, too, the swollen gelatinous capsules are fused together to make a gelatinous mass (**zoogloea** condition), in which the separate Bacteria are imbedded. This gelatinous mass forms the scum so frequently seen on the surface of putrescent fluids.

Multiplication takes place either by **spore formation** or by **fission** (hence their name of Fission Fungi); the daughter cells in the latter case either remain attached to their parents or are set free. Spore formation occurs usually, but not exclusively, when nourishment is scarce. In most cases the spores are formed endogenously; this occurs in many rod forms (Bacilli), and possibly in some other forms (Spirilla). When the rod has grown to a considerable length, bright highly refracting spores are seen within it; its contents now become granular and turbid, and it is at last dissolved, the spores being thus set free. Another method of spore formation has been described, with the formation of so-called arthrospores. This occurs in the following way:—if nourishment is scarce, one member of the chain becomes modified to form a rather larger, thicker walled

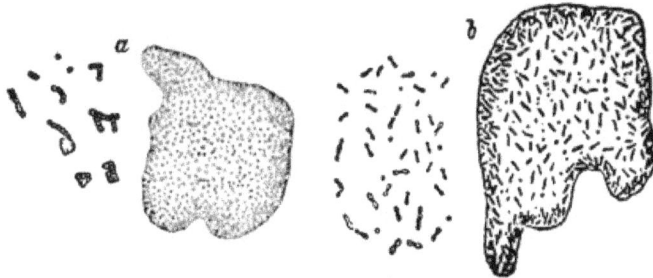

Fig. 99.—Bacteria (after Cohn). *a*, Micrococcus both isolated and in the zooglœa form; *b*, Bacterium termo, in free and zooglœa form.

cell than the rest; the remaining cells die and drop off, leaving the athrospore to build up a new colony.

Bacteria have been classed according to their shape into :—

1. **Micrococci**, round forms.

2. **Bacilli**, rod forms with a length two or three times as great as their breadth.

3. **Spirilla**, mostly motile, spirally twisted threads.

4. Forms with varying shape.

Such a broad classification is useful, but insufficient, owing to the very various results produced by the activities of different species with similar shapes. The separation of different species from one another is chiefly based upon (1) the differences in their mode of growth, (2) the different characters the colonies present when cultivated in different media, (3) the effects produced by their growth, (4) their relative size, and (5) the different ways in which they are affected by staining reagents.

In studying Bacteria, the following methods are usually adopted. A very small quantity of the material containing the Bacteria, is shaken up with some gelatine that has been made just sufficiently warm to be liquefied ; this is then spread out upon a plate, and after a time the gelatine will be seen to be dotted over with numerous small colonies of Bacteria. If these are examined microscopically, it can then readily be seen whether the colony examined consists of Bacilli, Micrococci, or Spirilla. The next step is to inoculate some nutrient material, such as gelatine, with a little of the pure culture obtained from the selected colony, and to note the rapidity with which the Bacteria grow, their mode of growth, their colour, whether the gelatine is liquefied or not, whether they grow down into it or remain on the surface, etc. Next the reactions of the Bacteria with regard to staining reagents are noted, and the final step with pathogenic Bacteria is to inject some of the pure culture into an animal, and to note the effects produced.

If some material which is supposed to contain the Bacillus anthracis, which gives rise to malignant pustule or anthrax, be taken as an example, it will be found that the following characteristic conditions occur :—

1. A small quantity of blood supposed to contain the Bacilli is shaken up with gelatine, and then spread out on a plate, which is then kept at a temperature of about 36° C. If the Bacilli are present, after twenty-four to forty-eight hours the colonies will appear as minute points.

2. These points are examined microscopically, and their behaviour with staining reagents is noticed.

3. Some of a colony which has been shown by the microscopic examination to be a pure culture of some Bacillus, is then transferred to nutrient jelly. After an interval of forty-eight hours, if the colony consists of the Bacillus anthracis, under a low power of the microscope it can be seen to be a small greenish-black spot with a wavy margin. If instead of the colony being placed on the surface of the nutrient jelly, it be plunged on the point of a needle into the jelly, the growth will be seen to consist of a delicate whitish line along the track left by the needle, from which fine threads run off at right angles.

4. At the end of two or three days it will seen that the jelly has com-

menced to liquefy in the neighbourhood of the growth, and after eight days it will be noticed that liquefaction has proceeded to a considerable extent.

5. It will further be noticed that the colony grows better the nearer it is to the surface of the jelly.

6. If now some of the pure culture be injected into the body of some animal liable to the disease, such as a rabbit or a mouse, anthrax will be invariably produced.

Owing to the minute size of Bacteria and their spores, these are constantly found floating about in the atmosphere. It is therefore absolutely necessary in all experiments with Bacteria to sterilise all the vessels, nutrient fluids, and other apparatus, before using them. This is usually done by exposing them to the prolonged action of heat, so as to kill any Bacteria or spores that may be adhering to them. After the apparatus and nutrient fluids have been sterilised, they must be carefully protected from accidental contamination.

The reason for making the first plate culture on gelatine, is to isolate the Bacterium that is being sought for from others that may be present, so as to obtain a culture which consists of only the one form that is to be studied.

As it is impossible to exclude air from the test-tubes containing the nutrient fluids, it is filtered through sterilised cotton wool.

Bacteria growing best in contact with the air, that is those that grow best on the surface of a nutrient substance, are called aërobic, whereas those that grow best in the absence of oxygen are said to be anaërobic.

CHAPTER XVII.

VORTICELLA.

THE Vorticella, or as it is also called the Bell animalcule, belongs to the same class as does the Amœba, that is to say, it is a **unicellular** animal. In the Amœba we have seen an animal consisting of protoplasm, in which the distinction between ectoplasm and endoplasm, though sufficiently clear to be seen, yet is not very well marked. In Vorticella we have an animal in which the separation of the protoplasm into these two layers is very definite; the **ectoplasm** forms a more or less tough external covering to the semi-fluid **endoplasm**. Going with this greater differentiation of the protoplasm we have of necessity various modifications of structure. As in the Amœba, the single cell has to perform all the functions necessary for the maintenance of life and for the reproduction of the species, but as the ectoplasm or ectosarc is tougher and more rigid, the animal can only change its shape to a limited degree, and is unable to throw out pseudopodia, for purposes of movement or of catching food, from any part of its surface. This inability to change its shape to any great extent, or to throw out pseudopodia, necessitates the development from one part of its surface of a large number of delicate short threads of protoplasm; these move constantly in the same direction, and therefore serve to set up currents in the water in which the animal lives. These currents are constantly flowing in such a direction that any small animals or plants which come under their influence, are swept into a sort of pit at one point of the surface of the animal. At the bottom of this pit the ectoplasm is less dense than at any other part of the surface, and consequently food particles can readily be forced, with a drop of water, into the endoplasm, there forming food vacuoles, as in the Amœba. The short delicate threads of protoplasm are called cilia ; each one is supposed to consist of two parts, the one side being elastic, and the other side contractile ; when the latter contracts, the thread is bent over to that side, then when the contraction passes off, the elasticity of the other side causes the cilium to become straight again. The result of the alternate contraction and relaxation of the contractile protoplasm is, that the cilium alternately rapidly bends and slowly straightens itself, and as the cilia all bend together and all in the same direction, the currents set up in the water are constant in their direction. Cilia react to the same stimuli as any other protoplasm. Moderate increase of temperature, weak chemical or electrical stimulation, etc., cause increase in their rate of movement ; whilst cold, withdrawal of oxygen, etc., retard it. So also a powerful electric shock, a strong chemical stimulus, or a temperature above 65° C. stops their movements altogether by killing them.

Vorticellæ occur both in fresh and salt water. There are different species of them, which vary considerably in size, the largest being just visible to the naked eye.

Anatomy.—The Vorticella is bell-shaped, hence its name of the Bell animalcule. The mouth of the bell, however, is not open, but is closed by a plate-like lid which fits inside its rim, only being apparently raised slightly at one side. This appearance is due to the bell being solid and its broad end being produced into a flange, which looks like the rim of the mouth of a bell, whilst the contents bulge into the space inside

12

the rim. This rim is called the **peristome**, and the part which

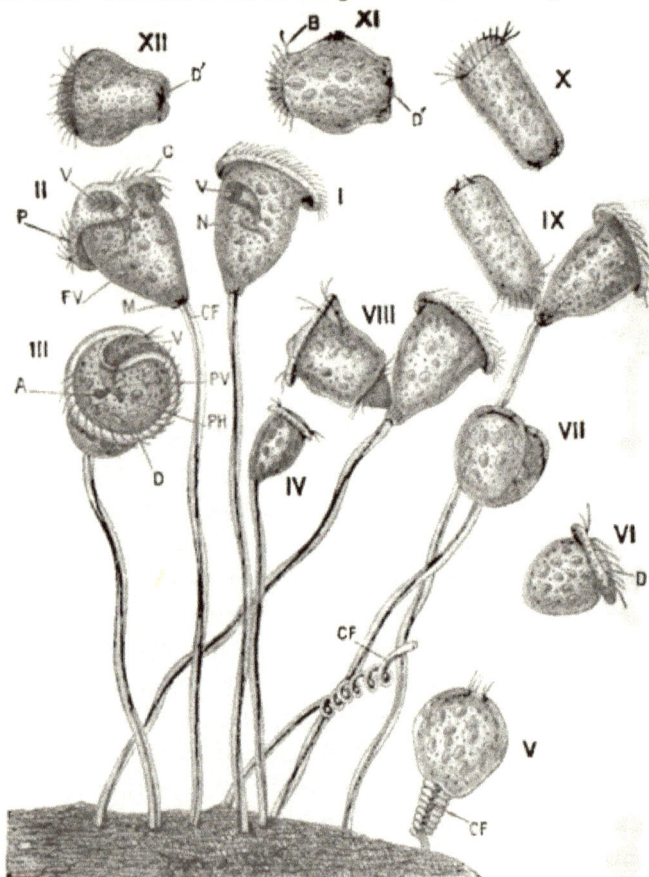

Fig. 100.—Vorticellæ (from Marshall and Hurst). I, II, III, IV, expanded; V, contracted; VI, separated from its stalk; VII, VIII, and IX show three stages of fission; X, XI, and XII show the separated individual swimming by means of the aboral circlet of cilia. *A*, food vacuole, discharging contents at anus; *C*, cilia; *CF*, contractile fibre of stalk; *D*, disc; *D'*, disc retracted; *FV*, food vacuole; *PH*, pharynx; *PV*, pulsating vacuole; *P*, peristome; *M*, myophan striation; *N*, nucleus; *V*, vestibule.

projects from the mouth and looks like a cover is described as the **disc.**

Between the disc and the peristome there is a fairly deep groove, and at one part this groove is hollowed out to form a pit, the entrance into which is described as the **vestibule.** The pit itself is called the **pharynx or mouth,** for in it occurs the weak spot in the ectosarc through which food particles find their way into the animal.

The part of this groove, which is on the opposite side to the pharynx, is not very distinctly marked.

There is a fringe of cilia running around the margin of the disc, and in the groove separating it from the peristome. They are also continued into the pharynx. The movements of the cilia are in such a direction that a stream of water is being constantly driven along the groove, and in and out of the pharynx, so that any solid particles in the water may drop down to the bottom of the pharynx, and thus be taken into the animal.

In the upper part of the pharynx there is another weak spot in the ectosarc, through which solid waste material, such as the remains of food, is discharged. This weak spot is described as the **Anus,** so that in this animal we have both a mouth and an anus.

The denser external layer of the protoplasm, that is, the ectosarc or ectoplasm, which covers the whole surface of the animal and lines the pharynx, is continued from the small end of the bell to form a slender **stalk,** by means of which the animal attaches itself to the stems of water plants, etc.

The external layer of this ectosarc is called the cuticle. It is very elastic and forms a protective covering for the animal, and is continued down on the stalk, forming the greater part of its thickness.

The deeper part of the ectosarc is contractile. At the lower part of the bell it presents the appearance of a number

of longitudinal lines or striæ, described as the myophan
striation.   These lines are continued down the centre of the
stalk as a darker band of contractile protoplasm.   When this
band of protoplasm contracts, the stalk presents the appearance
of a corkscrew.   As it relaxes itself, the elasticity of the cuticle,
which forms the remainder of the thickness of the stalk, is free
to act, and the stalk again straightens itself.   The alternate
contractions and relaxations of the contractile protoplasm are
not rhythmical, but only occur when the animal is touched by
some foreign body, or when the water in which it is living is
disturbed.   When this occurs, and the deeper part of the
protoplasm of the ectosarc contracts, the result is that both the
stalk contracts and the peristome becomes partially shut over
the retracted disc, so that the animal appears not unlike a
minute ball attached to the end of a corkscrew.

The endosarc, as in the Amœba, consists of a more fluid
granular protoplasm, which contains (circulating in it) **food
vacuoles** which get rid of their water so as to bring the food
into direct contact with the protoplasm, the water again collecting to form a single pulsating or contractile vacuole which is
situated between the disc and the vestibule, and which dis-
charges itself into the vestibule.

The **pulsating vacuole** acts as an excretory organ for getting
rid of water which holds in solution the soluble waste products.
The **nucleus** is a long horseshoe-shaped body lying in the
endosarc.   It is readily made plain by staining the animal with
magenta.   In addition to the nucleus a smaller body has been
described which lies close to the nucleus but not in it, and
the name paranucleus has been given to it.

Reproduction.—Three forms of reproduction are described
as occurring in the Vorticella.

1. **Fission.**—The animal grows broad, and divides in a

longitudinal direction into two parts, each of which contains half of the nucleus, half of the endosarc, and half of the ectosarc. The stalk, however, does not divide, but one half of the animal retains the whole of the stalk, whilst the other half develops close to its base a zone of cilia, called the aboral

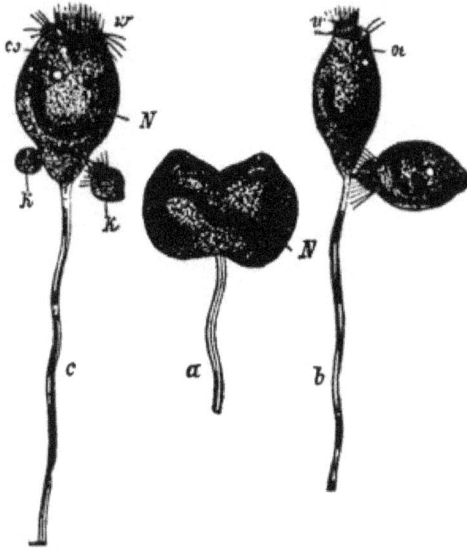

FIG. 101.—Vorticella microstoma (after Stein): *a*, in process of fission ; *N*, nucleus ; *a*, gullet: *b*, fission completed and aboral pole of cilia developed on free individual ; *w*, peristomial cilia : *c*, vorticella in process of conjugation; *K*, the small forms attached.

circlet of cilia. It then drops off the stalk and swims freely, its aboral circlet of cilia causing it to move in a rotatory manner through the water, until it strikes some object to which it can adhere by its base. It then develops a new stalk of its own.

2. **Conjugation.**—This does not appear to occur frequently. A small free-swimming Vorticella, formed by the rapid and repeated division of a large form, fuses with an ordinary large

stalked form.   The large Vorticella absorbs the small one into
itself, with the result that it seems to renew its youth, and is
capable of recommencing active division in the manner
described before.

3. The third form, which is called **Encystment with spore
formation,** is of rather doubtful occurrence.   The description
given is, that an ordinary stalked Vorticella, either after conjugation with a small free-swimming form has taken place or
without conjugation, becomes separated from its stalk, and
develops a firm, thick membrane around itself.   Then after
remaining quiet for some time, its protoplasm breaks up into
spores, which are round or oval bodies.   The enclosing membrane or cyst then ruptures, setting free the small bodies ; each
of these then develops a basal circlet of cilia, by means of
which it swims freely.   After a short free-swimming stage it
adheres to a piece of weed or other body, and then develops
a stalk and circlet of cilia around the disc, and grows up into a
new Vorticella.

CHAPTER XVIII.

GREGARINÆ.

G REGARINIDÆ belong to the Sporozoa, one of the classes into which the Protozoa have been divided. Leuckart defines them as "**unicellular parasites** of definite form, without pseudopodia or cilia, but covered by a smooth more or less firm cuticle." At their anterior ends there are not infrequently organs of attachment, like prosbosces or cushions. They are generally found associated together in large numbers (hence the name). Their movements are inconspicuous, wormlike, or slightly amœboid. They live wholly as parasites, and receive their food by endosmosis. In some animals they are found in almost every specimen examined. Sometimes they occur in the body-cavity, sometimes in the intestine, the largest known Gregarina having its home in the rectum of the Lobster, and sometimes in other organs, as for instance, the seminal vesicles of the Earth-worm.

At present they are only known to occur in Invertebrates. They especially infest Crustaceans, Insects, and Worms.

Closely allied forms, the Coccidia and the organisms known by the name of Psorosperms, occur in large numbers in Mammalia.

The family of the Gregarinidæ is divided into two main groups, the Monocystidia in which the cell is undivided, and the Polycystidia, in which it is divided into two chambers. The former chiefly occur in Worms, the latter chiefly in Insects, Lobsters, and Crayfish.

The Gregarina lives in a very protected situation, either within the intestine or in some other organ of the animal upon which it is parasitic. It is, therefore, unnecessary for it to develop a rigid resistant ectosarc to serve as a protective covering. The result of the less rigid nature of the ectosarc is that the animal can readily change its shape. Again, owing to there being no necessity for a very resistant cuticle, fluids can be readily absorbed by the endosarc through the ectosarc. The animal obtains all its nourishment in a liquid form, for its host, by means of its digestive juices, converts all the food it obtains into a rich nourishing liquid. The Gregarina lives in this liquid, and absorbs it through its walls.

This mode of life renders a mouth unnecessary, and in consequence the animal does not possess one. Reproduction takes place by means of more or less hard-shelled spores (pseudonavicellæ), which are formed in the interior of the adult, sometimes in very great numbers. Sooner or later, there generally develop in these spores small sickle-shaped bodies, which creep out and become new parasites. The number of these varies, but usually there are not many of them ; occasionally the contents of the spore are collected in a single amœboid embryo.

Anatomy.—The body of the Gregarina is more or less extended. The protoplasm, of which it consists, is divided into two parts, a clear external layer, the **ectosarc**, which is provided with a thin cuticle ; and a granular internal portion, the **endosarc**, in which there is a large **nucleus**, containing a nucleolus. The deeper portions of the ectosarc occasionally present a distinct striation, similar to the myophan striation in the Vorticella. It has been stated that the granules in the endoplasm sometimes give a starch-like reaction with iodine. Occasionally the endoplasm is coloured yellow with changed blood-pigments which the animal has absorbed from its host. No contractile vacuole is ever present ; occasionally, though rarely, one or more small vacuoles are seen.

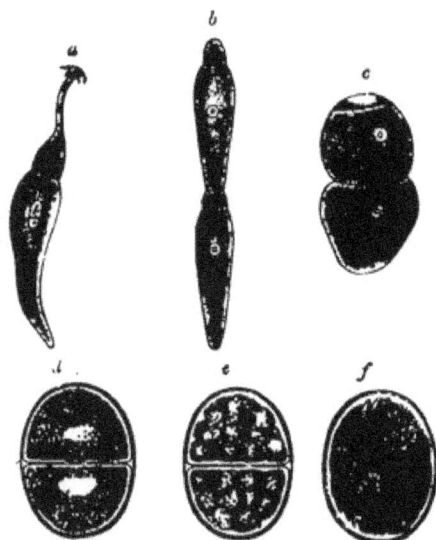

FIG. 102.—Gregarina (after Stein and Kölliker). a, Gregarina oligacanthus from intestine of a Dragon fly ; b, Gregarina polymorpha from the intestine of a Meal beetle during conjugation ; c, in process of encystment ; d, encysted Gregarina; e, formation of pseudonavicellæ ; f, pseudonavicella cyst with ripe pseudonavicellæ.

Those forms inhabiting the intestines are often provided with hooks at one end, and sometimes with suckers, by means of which they cling to the intestinal walls of their host. These hooks do not persist through life, but fall off sooner or later.

It is not rare to find two Gregarinæ attached together by their suckers. This sometimes occurs in young forms, and may be quite independent of the pairing which takes place during the process of conjugation.

The two-chambered form differs in no essential particular from the one-chambered, for though the animal has been divided by an ingrowth of the ectosarc into two chambers, yet it is still unicellular, the nucleus remaining unchanged. The partition between the two chambers is only a septum, which has developed as the animal approached adult age. The larger or posterior chamber retains the nucleus. The formation of this septum never indicates approaching fission, nor indeed is it known to have any special significance. It only occurs in the forms which inhabit the alimentary canal.

Reproduction and Life History.—Two, or occasionally three or more, adult forms may unite together end to end with their long axes lying in the same straight line. If reproduction is going to take place, this will be the first stage in a process of conjugation. The Gregarinæ soon contract and surround themselves with a gelatinous wall not unlike cellulose in appearance, but of quite different chemical composition. This gelatinous wall is called a cyst, and the process is described as **encystment** (fig. 102). Whilst this encystment is taking place the individuals become completely fused together. Occasionally a single Gregarinæ will encyst itself and go through the stages of reproduction exactly in the same way as though conjugation had taken place. The contents of the cyst now rapidly divide (exactly how has not yet been observed), into a number of small ovoid spore-like bodies. These bodies are probably nucleated, but so far no nuclei have been proved to exist. The spore-like bodies soon elongate into spindle-shaped organisms, each one developing around itself a horny, colourless coat. These spindle-shaped bodies are called **Pseudonavicellæ**; the cyst is described as a pseudo-navicella cyst. After a time the cyst ruptures, and the pseudonavicellæ are set free. The protoplasm of each pseudonavicella becomes divided up into several small sickle-shaped bodies, each of which is called a **falciform body** or corpuscle. Later on these falciform bodies are set free from the pseudonavicella by the rupture of its horny coat ; each one is then generally found to be provided with a short vibratile process by means of which it moves about until it reaches a cell of its host. It then makes its way into the cell, and takes up its abode in it. Within this cell it develops at the expense of the protoplasm until it is able to take up nourishment from the nutrient fluids of the host which bathe the cell. It thus increases rapidly in size, stretching and eventually rupturing the cell that has given it house room. It may now remain in the same place, or it may be carried to some other organ of the body from that in which it has developed. In either case it undergoes further development until it reaches the form and appearance of the adult Gregarina.

It is not improbable, that in some cases the pseudonavicella cyst escapes from the host, wherein the adult Gregarina has lived, in the intestinal forms with the fæces, or in the forms inhabiting the seminal vesicle of the earth-worm, either by the death of the worm or possibly after rupture of the pseudonavicella cyst. The pseudonavicellæ may pass down the male genital ducts of their host, and thus get to the exterior. The pseudonavicellæ, or sometimes the whole cyst, may be eaten by another animal, in which rupture or digestion of the horny coat of the pseudonavicella may take place, with the result that the falciform bodies are set free ; they would then bore their way into the cells of this their intermediate host and there develop as before described, finding their way into the final host after the death of the intermediate host. The above life history of intermediate and final hosts is not improbably a true one, but it has not been proved to occur, and it seems undoubtedly the case that the whole of the life history can be passed in a single animal. The life history of the Gregarinæ presents many points that have not yet been conclusively decided ; the difficulty of following out the various processes is very great, and the observations that have been made by different observers do not always agree.

CHAPTER XIX.

HYDRA.

THE animals so far studied have all been unicellular. In them the single cell has to perform all the necessary functions for the maintenance of life and for the reproduction of the species. There has been differentiation, but it has only been differentiation of protoplasm. Thus the ectosarc serves as a protective covering, and in some cases, as in the Vorticella, can be subdivided into a contractile portion and a cuticular portion. The endosarc, on the other hand, digests the food, and gets rid of waste products, etc. This differentiation is of the greatest importance, but it is evident that, if the animal were to increase much in size and complexity, it would be manifestly to its advantage that instead of consisting of one cell, the protoplasm of which was very much diffentiated, it should consist of many cells, the cells themselves being differentiated off into groups, and the work of the whole animal being divided into various parts, each part being performed by a group of cells. This, of course, is only repeating what has been already said in the description of tissues.

The Hydra is a **multicellular animal**, in which the differentiation of protoplasm and of cells has only taken place to a limited extent, but quite sufficiently to place it on a much higher level than the Amœba, the Vorticella, or the Gregarina.

In the Hydra the cells are arranged in two layers, an outer
layer or skin called the **ectoderm**, and an inner layer called the

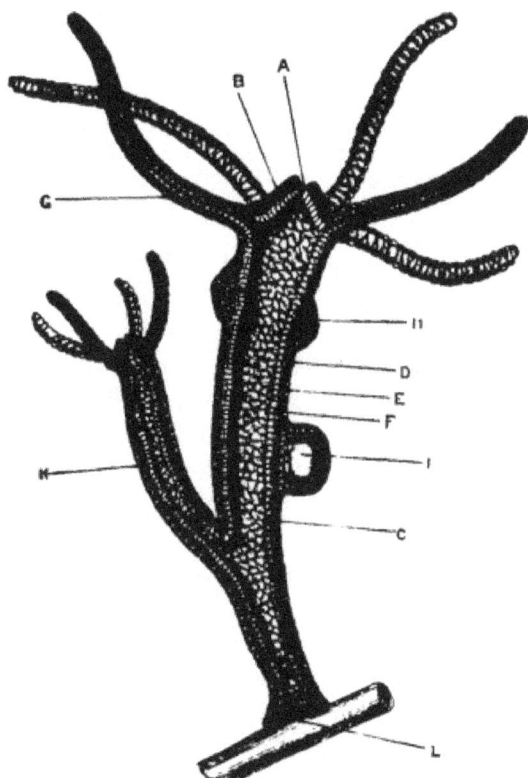

FIG. 103.—Hydra; a diagrammatic longitudinal section of a specimen with ripe
reproductive organs, and with a fully formed bud (from Marshall). *A*, mouth;
B, hypostome; *C*, enteron or digestive cavity; *D*, ectoderm; *E*, supporting
lamella; *F*, endoderm; *G*, cavity of one of the tentacles; *H*, testis; *I*, ripe
ovum in the ovary; *K*, fully formed bud with mouth and tentacles; *L*, root by
which the Hydra is attached to a piece of weed.

endoderm. These terms, ectoderm and endoderm, the inner
and outer skins, must not be confused with the terms
ectoplasm and endoplasm, the inner and outer layers of

protoplasm in the unicellular animals. This arrangement of
cells forms a two-layered sac, which has a mouth at one end
opening to the exterior. This sac is the **primitive alimentary
canal or enteron.**

The adult Hydra bears a close similarity to the higher animals in the
diblastula or gastrula phase, described under Development. The animals
like the Hydra, in which this diblastula condition is the ultimate one,
development proceeding no farther, are grouped together under the name
of the Cœlenterata.

The Hydra is a fresh-water animal, readily seen with the
naked eye. It is occasionally found in ponds and ditches,
attached by its base, or so-called foot, to stones or weeds. It
prefers, however, to live in shallow, clear streams, where the
water is not flowing too rapidly. Three species commonly
occur in England, the Hydra viridis, distinguished by its green
colour and small size, the Hydra fusca which is brown, and
the Hydra vulgaris which is almost colourless.

Anatomy.—The animal consists of a tubular body, the wall
of which is composed of two layers of cells. The cavity or
enteron is closed at one end but open at the other. The
opening is called **the mouth.** It can readily be distended for
taking in food. Around the mouth the body wall is produced
into six or eight blunt processes or **tentacles**, each of which
consists of a wall and a cavity, which are the direct continuation
of the body wall and the digestive cavity. When fully extended they may be twice the length of the body, whilst when
contracted they appear only as short, blunt projections around
the mouth. The so-called foot is the closed end of the body.
It is somewhat club-shaped, and keeps a firm hold on the
stone or other object to which the Hydra is attached, by
spreading itself out on its surface. The animal can alternately
contract and relax this foot, thereby giving rise to a slow

crawling movement. Around the mouth, forming a circular lip to it, is **the hypostome**.

Minute structure.—If a thin transverse or longitudinal section be made through the body wall of a Hydra, a microscopical examination can readily be made of the cells of which

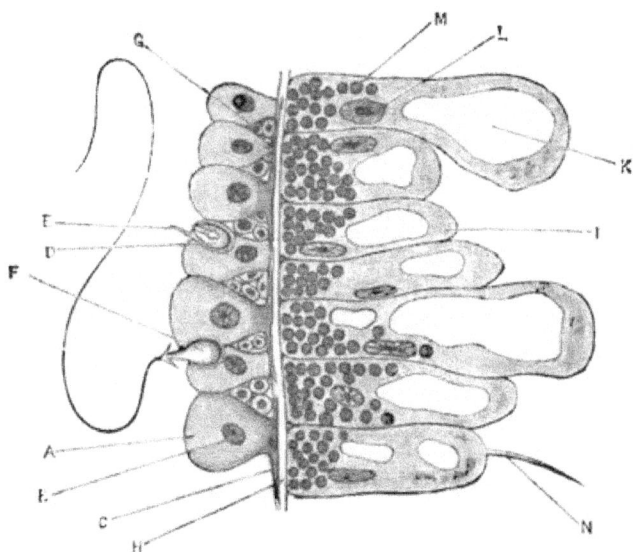

FIG. 104.—Hydra viridis; a small portion of a longitudinal section through the body-wall (from Marshall). *A*, one of the large ectoderm cells; *B*, nucleus of the same; *C*, muscular process of the same; *D*, an undischarged nematocyst lying within its parent cell or cnidoblast; *E*, cnidocil; *F*, a nematocyst with discharged thread; *G*, interstitial cell; *H*, supporting lamella; *I*, endoderm cell; *K*, vacuole; *L*, nucleus, *N*, flagellum of same; *M*, chlorophyll corpuscle.

the animal is composed. The whole thickness of the body-wall may be seen to consist of three parts—an outer layer of cells, the **ectoderm**; an inner layer of cells lining the digestive cavity, the **endoderm**; and a third, very thin gelatinous layer, which can only be seen with some difficulty owing to its transparency, and which separates the ectoderm from the endoderm, called the **structureless** or **supporting lamella** or mesoglœa. The ectoderm consists of four kinds of cells which differ in their functions, and have a corresponding difference in shape, size, and appearance. The large ectoderm cells form a single layer covering the whole surface of the body, including the tentacles. It will be observed from the figure (fig. 104) that each of these large cells touches its neighbour by its wide ends. The surface formed by the upper ends is developed into a horny cuticle. The lower ends of the cells form muscular processes which are spread out on the structureless lamella. In the spaces left between the bodies of the large cells, numerous small rounded cells (interstitial cells) are packed. Amongst these there are a large number which have become specialised; they are called cnidoblasts, and in each there is a thread cell or nematocyst, which contains a spirally wound thread. The cnidoblast, after the formation of the thread cell, remains as a capsule to it. There is a small projection (cnidocil) from it, which stretches beyond the free surface of the animal.

If the cnidocil be touched, the thread cell at once throws out its thread, which is turned inside out (evaginated). The thread is coated with some irritating fluid, and thus the animal is provided with a kind of weapon, which it can shoot out at an enemy or at prey. Sometimes the whole thread cell is shot out; it is able to stick on to the enemy, as it is provided with barbs at the base of the thread. Each nematocyst possesses a well-marked nucleus.

In addition to the above kinds of cells some star-shaped ganglionic cells occur. These are nervous in function, and are in special relationship, by means of fine connections, with the thread cells, and with the contractile muscular processes of the large ectoderm cells.

The endoderm consists of a single layer of large cells with well-marked vacuoles. Processes, somewhat similar to the pseudopodia of the Amœba, are sent out by these cells into the digestive cavity, and food particles are taken in by them, exactly as in the Amœba. The endoderm cells also bear permanent processes, which wave about in the digestive cavity. These are similar to cilia, only they are larger. They are called flagella. Muscular processes to the endoderm cells have also been described ; they are said to run transversely around the body as opposed to the longitudinal direction of the muscular processes of the ectoderm cells. In the Hydra viridis, the endoderm cells contain numerous chlorophyll corpuscles, which resemble the chlorophyll corpuscles in plants, in having the same power of absorbing and decomposing carbon dioxide and of setting free oxygen in the presence of sunlight. In the Hydra fusca similar bodies are found, but they contain no chlorophyll.

Both the tentacles and the body of the Hydra are extremely contractile. The animal, if frightened or injured, can contract up into a very small compass. This is brought about by means of the muscular processes of the large ectoderm cells. As before mentioned the animal can crawl slowly by means of its foot, but it can move much more rapidly, by fixing its mouth, contracting the body, then fixing the foot, letting go the mouth, extending the body, fixing the mouth again, contracting the body, and so on.

The Hydra feeds chiefly upon other animals. By the constant and rapid movements of its tentacles it can search for its food over a considerable area. When it meets with an animal that it can attack, it paralyses it by discharging a volley of thread cells at it. Then it takes it in through its distensible mouth, and digests it at leisure.

13

Digestion takes place in the Hydra in two ways: (1) As in the Amœba, by taking particles of food directly into the endoderm cells, *i.e.* by **intracellular digestion**, and (2) by acting on the food in the enteron, by pouring upon it digestive juices, and when digestion has taken place, that is, when the food is rendered soluble, by absorbing this solution. This is called **intercellular digestion**, or digestion between or outside the cells. It is necessary, however, in intercellular digestion, that the food should come into contact with the surface of the endoderm cells.

Reproduction.—The Hydra reproduces itself in two ways, **asexually** and **sexually.**

Asexual reproduction or **gemmation.**—This consists in the outgrowth, from some part of the body, of a hollow bud, which first acquires a mouth and tentacles, and then becomes constricted at the base, in order to shut off communication between the enteron of the child and that of the parent. Two or three buds may develop at the same time from the same animal.

Another mode of asexual reproduction has been artificially induced. The animal was cut up into several pieces, and it was observed that those which contained both ectoderm and endoderm grew into perfect Hydræ. This mode of reproduction by fission is not known to occur naturally.

Sexual reproduction.—The Hydra is hermaphrodite, the same individual bearing both male and female organs. There may be numerous testes on the same animal, and sometimes, but more rarely, several ovaries (Hydra fusca). Both ovaries and testes are very simple, and apparently self-fertilisation can take place. In this the Hydra is unlike most hermaphrodite animals, in which self-fertilisation is not possible, there being various contrivances to prevent its occurrence.

Both the testis and the ovary develop from groups of interstitial cells. In the case of the testis, a group of interstitial cells multiplies, until it has formed a heap projecting from the surface of the animal. The outer cells then take on a protective function, whilst the cells in the interior of the heap become directly transformed into spermatozoa. These escape when ripe, the nuclei forming the heads, and the protoplasm the long tails, by the lashing movements of which the spermatozoa make their way through the water to the egg cell.

An ovary develops in a similar manner by the multiplication of a group of interstitial cells; but whereas in the testis all the central cells develop into spermatozoa, in the ovary it is only one of the central cells that develops into an ovum or egg cell. This one grows at the expense of the rest, until it entirely fills up the space within the outer protecting cells. As soon as the ovum is quite ripe, the capsule formed by the protecting cells ruptures, the polar bodies are extruded, and through the rupture a spermatozoon makes its way into the egg cell, which thus becomes fertilised.

The Development of the Hydra.—After fertilisation, segmentation at once begins; the ovum surrounds itself with a hard thick capsule, and, becoming detached from the parent, falls to the bottom of the pond in which the animal is living.

The further stages of development have been differently described by different observers. According to one authority, as the result of segmentation, a solid morula is formed, consisting of a single layer of cells, each of which divides into two, so as to form two layers, the ectoderm and endoderm. Next, a cavity develops in the middle of the mass, and thus a closed diblastula or gastrula is formed. If this be the true account, the development would be by pure delamination, the cavity, which develops in the centre of the mass, being only an increase in size of the already existing, but minute, segmentation cavity. According to another observer, the segmentation of the ovum results in the formation of a hollow morula or blastosphere, with a well-developed segmentation cavity. From one part of this blastosphere, cells are budded off which migrate inwards, and fill up the segmentation cavity, thus forming a more or less solid mass inside the original layer. These cells ultimately form the endoderm. In the centre a cavity develops, which is the enteron or alimentary canal. This mode of formation of the diblastula is a modified form of delamination, and is called the immigration form of delamination. Whether the former or the latter description is correct, the result is the same; namely, that a closed two-layered sac is formed, which is surrounded by its chitinoid (capsule secreted by the ectoderm cells); this capsule soon bursts, and the embryo escapes; it then elongates itself, the sac becomes ruptured at one end to form the mouth, and around the mouth, tentacles bud out. Thus a young Hydra similar to the parent is formed.

CHAPTER XX.

THE LIVER-FLUKE.

THE Flukes are **parasitic unsegmented flat Worms**, with leaf-shaped bodies; they possess a **mouth** and two ventrally placed **suckers** or organs of attachment. The **intestine** is forked and **without an anus**. The **Liver-fluke** (*Distomum hepaticum*) is very frequently found in the liver of the Sheep, being sometimes present in large numbers in the bile ducts. It produces the disease known as sheep rot. It may occur in Man, and in many other animals, and is not confined to the liver, though that is the place where it is most frequently found. At one stage of its existence, it inhabits the Pond-snail. The adult animal is about $1\frac{1}{2}$ inches long.

In the Hydra we have seen a multicellular animal, which has only a simple structure. It consists of only two layers of cells; it has at most the merest trace of a nervous system, no special organs for getting rid of waste products, a very simple digestive cavity, and very simple reproductive organs. In the Fluke, we have to study an animal in which all the above systems are complicated; that is, an animal in which differentiation has taken place to a much greater extent than it has in the Hydra. The complicated nature of the anatomy and the life history of the Fluke is, however, largely due to the fact that it is a parasite. Its mode of existence renders some organs unnecessary, so that they tend to disappear; but on the other hand the difficulties of maintaining the species are so great, that it is provided with a most complex generative system, and the young pass through various phases before they reach maturity.

Anatomy.—The adult Liver-fluke is about $1\frac{1}{2}$ inches long and $\frac{1}{2}$ an inch broad at its widest part. It is of a flattened ovoid shape, and is much broader at its anterior than at its posterior end. There is a blunt projection from its anterior end which is roughly triangular in shape, and at the end of this there is a cup-shaped **sucker**, in the middle of which the oval **mouth** is placed.

The **ventral sucker** or second organ of attachment is also cup shaped; it lies on the ventral surface just behind the junction of the triangular projection with the broader portion of the body. Both suckers are well supplied with strands of muscle, which enable them to hold on with considerable tenacity to anything to which they may attach themselves.

The whole surface of the body is covered with a **cuticle,** which is provided with numerous spines.

The Alimentary system.—The **alimentary canal** consists of a **mouth**, a **pharynx**, an **œsophagus**, and an **intestine**. The mouth is oval, and is placed in the middle of the anterior sucker. The pharynx is oval, and its walls are well provided with muscles. It leads into the œsophagus, which is a very short, straight, thin-walled tube. This again leads into the intestine, which is a forked tube ending blindly. The intestine divides almost immediately into the two forks, one on the right, and one on the left, each of which runs to the hinder end of the body, giving off numerous short branches on its inner side, and many large branches, which subdivide again into smaller branches, on the outer side. None of the branches open to the exterior, nor, having once left the main trunk, do they

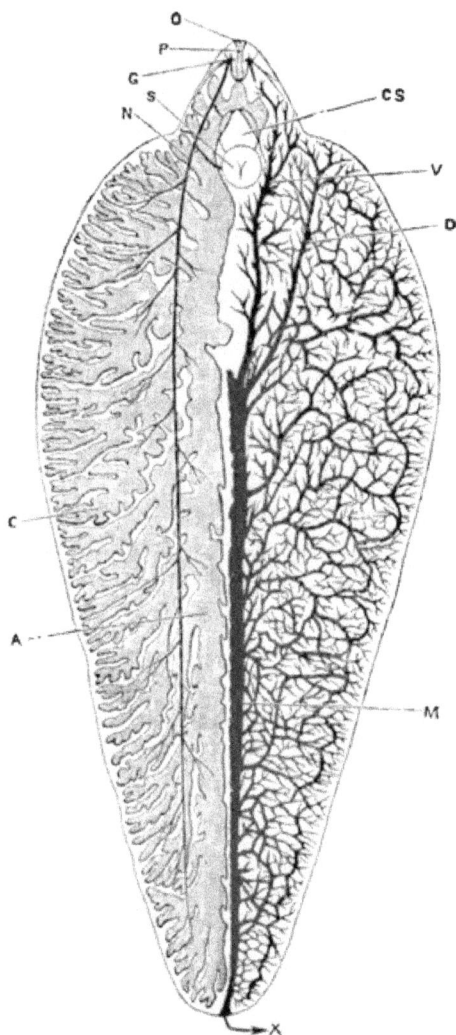

FIG. 105.—Distomum hepaticum (from Marshall after Sommer). The alimentary and nervous systems are shown only on the left side, and the excretory system only on the right. *A*, right main division of the alimentary canal; *C*, one of its diverticula; *CS*, cirrus sac; *D*, left anterior dorsal trunk of excretory system; *G*, lateral ganglion; *M*, main trunk of excretory system; *N*, lateral nerve; *O*, mouth; *P*, pharynx; *S*, ventral sucker; *V*, left anterior ventral trunk of excretory system; *X*, excretory pore.

communicate with each other. There is no anus, nor has the digestive system any glands connected with it.

The Excretory system.—This consists of a complicated system of branched **ducts**, which open internally to the body-cavity, and externally by a **median pore.** The small-branched ducts commence by **ciliated funnels**, which open into the body-cavity. The ducts anastomose freely with one another, and finally open into longer **transverse ducts**, which open into a tube of considerable size, called the **main duct.** This is formed by the union of four **anterior ducts**, a pair of dorsal, and a pair of ventral, which unite just opposite the broadest part of the animal, and run from this point directly backwards to open to the exterior by the terminal median pore.

The body-cavity, or cœlom, into which the ciliated funnels open, is represented by a series of spaces or lacunæ, which occur between the branched alimentary canal and the body wall.

The branched condition of the **alimentary canal** allows nourishment to be readily carried to all regions of the body, and so does away with the necessity of any special apparatus for the distribution of food. Further, as the animal is parasitic within other animals, it gets its food in a liquid form; hence no biting or masticating organs are required. Again all its food is digestible, so that there are no solid residues to get rid of, and hence the animal has no need of an anus. Soluble waste products, produced by the activity of the protoplasm of the cells, there are in plenty, and these are got rid of by means of the excretory system.

The **Nervous system** can only be seen with some difficulty. It consists of a nerve collar which surrounds the pharynx, and which has upon it three collections of nerve cells or **ganglia.** One of the ganglia is on the ventral surface of the **nerve collar,** and short nerve branches arise from it to supply the surrounding organs. The other two ganglia (the **lateral ganglia**) occur one on each side of the nerve collar; from each of these there is a long lateral nerve, which extends to the posterior end of the body, giving off numerous branches on its course to the various parts.

Reproductive system.—The **genital aperture** is on the ventral surface of the animal between the two suckers. The reproductive system, as in most parasites, is very complicated, the complexity being still further increased by male and female organs both being present in the same individual, that is, by the animal being **hermaphrodite.**

The **Male** organs consist of two much-branched **testes**, which lie, one behind the other, in the middle part of the body, extending over about half its length and half its width. Each testis has, arising from about its middle, a tube, the **vas deferens.** The two vasa deferentia run forward as far as the ventral sucker, where they open into an elongated sac, the **vesicula seminalis.** Opening out of the seminal vesicle and forming a narrow continuation of it, there is a fine tube, the **ejaculatory duct,** which runs to the end of the **penis.** The penis is a well-developed, muscular organ, forming, as it were, an enlarged continuation of the ejaculatory duct; it can be protruded by a process of evagination from the genital pore. The vesicula seminalis, the ductus ejaculatorius, with a small **accessory gland** which surrounds it, and the penis when withdrawn, all lie within a sac-like space between the ventral sucker and the genital aperture. This is called the **cirrus-sac.**

The **Female organs** are still more complicated. At the base of the penis, when it is withdrawn, there is a small space called the **genital sinus,** into the left side of which the **oviduct** opens. If the oviduct be traced backwards, it will be found to be a thick-walled, much-convoluted tube, lying between the genital aperture and the commencement of the testes. Its convolutions extend halfway across the animal,

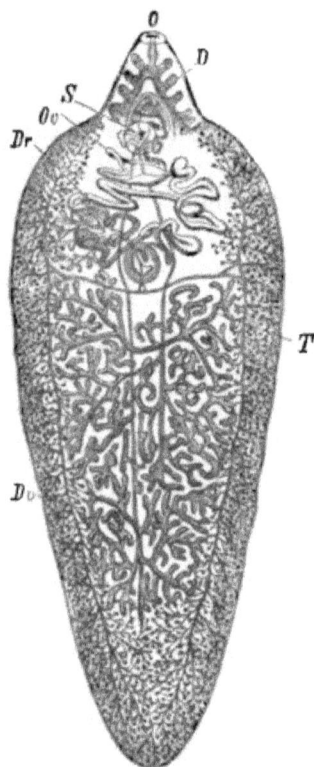

FIG. 106.—Distomum hepaticum (after Sommer). *O*, mouth; *D*, limb of intestine; *S*, sucker; *T*, testes; *Do*, yolk gland; *Ov*, uterus; *Dr*, accessory glands.

and it may be seen to contain numerous eggs. It is often called the **uterus**. At its termination it receives two ducts, one from the **ovary**, the other from the **yolk glands**.

The ovary is branched and tubular, looking not unlike the

testis but smaller; it lies on the right side in front of the anterior testis, and the various branches unite together to form a fine tube, the ovarian duct, which opens into the uterus (oviduct). The Shell gland lies in the middle line. It is a glandular mass, which surrounds the termination of the ducts from the ovary, and also the median duct from the yolk glands, just at the point where they unite to form the uterus or oviduct.

The Yolk glands consist of a very large number of small rounded masses, which lie on either side of the body, extending along its whole length. Each collection of yolk glands extends from the outer margin inwards for a quarter of the width of the worm.

The Vitellarian ducts.—Each yolk gland has connected with it a fine duct (a vitellarian duct). These ducts all unite to form an anterior and a posterior duct on either side. The anterior and posterior ducts unite about the junction of the anterior and middle thirds of the animal, to form a transverse duct which runs inwards and opens into a small pouch, called the **yolk reservoir.** From this a short duct, the median vitellarian duct, runs, which joins with the ovarian duct to form the uterus or oviduct. Just at the junction of the median vitellarian duct and the uterus, there is a small duct, which is said to open to the exterior on the dorsal surface of the animal, and is called the **vagina** or "canal of Laurer."

Life history.—The adult Liver-fluke, having developed eggs in its ovary, passes them into the ovarian duct. They are probably fertilised whilst on their way to the uterus, either by spermatozoa from the same animal or from another animal. After fertilisation the eggs pass into the uterus; they are there supplied with yolk cells from the yolk glands, and become encased in **horny shells** by the shell gland. The eggs then

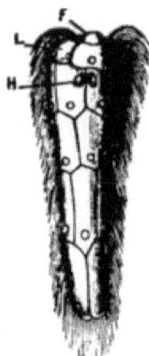

FIG. 15
× 250

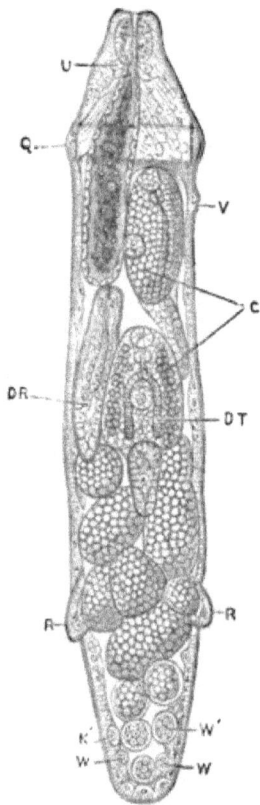

FIG. 18
× 70

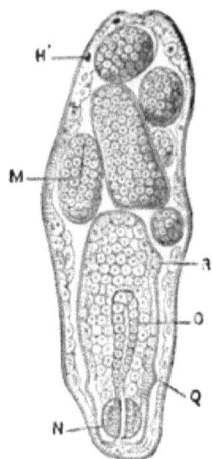

FIG. 16
× 115

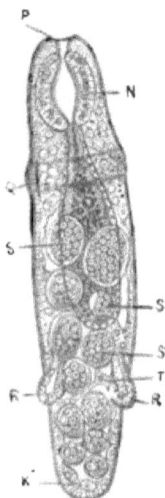

FIG. 17
× 115

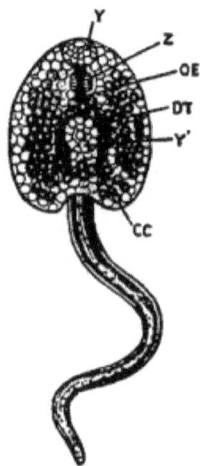

FIG. 19
× 80

FIG. 107.—Fasciola hepatica. Five stages in the life history (Marshall and Hurst, after Thomas). Fig. 15, the free swimming embryo; fig. 16, a sporocyst containing developing rediæ; fig. 17, a young redia; fig. 18, an adult redia containing one daughter redia (*DR*), two cercariæ nearly mature (*C*), and germs

grow rapidly in size and are passed out of the body by the aperture at the base of the penis, and are thus ejected into the bile ducts of the sheep or other animal in which the Fluke is living. They then pass through the bile ducts with the bile into the intestine of the host, and then escape from the body with the fæces.

Each egg is an ovoid body, ·13 mm. long and ·08 mm. broad, enclosed in a smooth brown horny shell, and containing a **single germ cell** and **numerous yolk cells**. As soon as the egg is laid, active development commences, and proceeds rapidly until an embryo is formed within the shell. At the same time one end of the shell becomes separated from the rest to form an **operculum** or lid, and it is through the opening thus formed, that the embryo escapes. The **free-swimming embryo** is conical in shape; it can swim rapidly by means of the cilia covering its body; its **ectoderm** consists of a single layer of flattened cells enclosing a mass of granular ones, and it has two eye spots, a head papilla, and a pair of ciliated funnels. Usually it lives for only about eight hours; if during this time it is so fortunate as to meet with a small pond-snail (the intermediate host to the Fluke) the head papilla will elongate, and the animal will bore its way into the snail's body. As soon as it arrives at a suitable resting-place, such as the pulmonary chamber, changes at once begin to take place, the cilia disappear, the ectoderm cells swell, and after an interval of two

---

in various stages; fig. 19, a free cercaria. *CC,* cytogenous cells of cercaria; *DT,* the digestive tract; *F,* head papilla; *H,* eye spots; *H′,* the same degenerating; *K′,* germinal cell; *L,* epaulet-like cells of·first row of ectoderm cells; *M,* embryo in optical section, gastrula stage; *N,* pharynx of redia; *O,* digestive sac; *OE,* œsophagus of cercaria; *P,* lips of redia; *Q,* collar of redia; *R,* processes of redia serving as rudimentary feet; *S,* embryos between the morula and gastrula stages; *T,* trabeculæ. crossing body-cavity of redia; *U,* cells in redia supposed to be glandular; *V,* birth opening by which cercariæ and daughter rediæ escape from the redia; *W,* morula still retained in the body-wall; *W′,* larger morula; *Y,* oral sucker of cercaria; *Y′,* ventral sucker; *Z,* pharynx of cercaria.

or three weeks the embryo has grown into an elongated sac
·6 mm. in length.

Sporocyst.—This elongated sac is called the **sporocyst**;
occasionally it divides into two similar sporocysts, but this
only rarely occurs.

So far, we have seen that, as the result of the sexual act,
fertilised eggs are produced, which, under suitable conditions,
develop into sporocysts. The sporocyst now gives rise, by
internal budding, to a number of young animals, called **rediæ**,
and thus a second (asexual) generation is produced. These
pass through a **solid morula** and a **gastrula** stage whilst still
within the sporocyst. Later on they burst their way through
the wall of the sporocyst, and then develop into adults, becoming elongated and provided with alimentary canals, in the form
of short blind sacs. After a redia has made its way out of
the sporocyst, the opening closes and the sporocyst goes on producing others. The adult redia, in addition to the alimentary
canal, possesses definite excretory canals, which commence in
the body-cavity as ciliated funnels. From the internal surface
of the body-wall new buds are given off, which generally
develop into new rediæ. Thus there may be several generations, but finally and sometimes without the first redia producing rediæ at all, **cercariæ** are produced. To recapitulate
the sporocyst gives rise to rediæ, which may in their turn
simply produce rediæ over again, but finally cercariæ are
produced. The cercaria develops exactly like a redia up to
the gastrula stage, but after that, whilst still within the body of
the parent redia, it becomes very different. It develops a mouth,
a pharynx, and an œsophagus; and below these, a forked
alimentary canal; a long tail, and two suckers also make their
appearance. These fully-formed cercariæ now escape from
the redia by an aperture just behind the collar; they wriggle

themselves out of the Snail into the water or on to the grass, and attach themselves to any object, such as grass stalks, etc., near which they happen to be. Each cercaria then soon loses its tail, and surrounds itself with a membrane; that is, it encysts itself. If the encysted cercaria is then eaten by a Sheep or other animal, fit to form its final host, it resumes its activity, emerging from its cyst, and passing along the bile ducts of the animal until it reaches the liver, where it takes up its permanent abode. In about six weeks it is mature.

Thus we see that the sexually mature Fluke gives rise to eggs, which develop into sporocysts. These produce rediæ asexually; the rediæ eventually produce cercariæ still asexually, and the cercariæ under favourable conditions grow up into adult Flukes. There is, therefore, a sexual generation followed by two or more asexual ones, the last of which is again followed by a sexual one.

The disease of liver rot in Sheep is a common one, and is often very fatal to large numbers of the affected animals; it has been known to destroy as many as a million sheep in one year in Great Britain alone.

CHAPTER XXI.

TAPE-WORMS.

THIS group forms one of the three divisions of the Platy-helminthes, the Trematodes forming the second, and the Turbellaria the third.

Leuckart defines the Tape-worms as "**flat worms without mouth or alimentary canal,** which typically develop by alternation of generations, by budding from a generally pear-shaped **nurse or scolex,** with which they remain united for a lengthened period as a ribbon-like colony or **strobila.** The individual joints of the colony, *i.e.* the sexual animals or **proglottides,** increase in size and maturity, as they are removed farther from their origin by the intercalation of new buds, but are not distinguished in any special way. The nurse, however, known by the name of the head (scolex), is provided with four or two suckers, and usually with curved claw-like hooks. The dorsal and ventral surfaces of the head are perfectly identical, so that the arrangement of the hooks presents a strikingly radiate appearance. By means of this apparatus, the worms fasten themselves on to the intestinal membrane of their hosts, which with only one exception, all belong to the Vertebrata. The nurses develop from little round six-hooked embryos in a more or less complicated fashion, as so-called bladder-worms. The latter inhabit very diverse, but usually parenchymatous, organs of the higher and lower animals, and are thence passively transferred to the intestine of their subsequent host."

The simplest forms of Tape-worms possess only one segment or so-called joint, and are furnished with only one set of (hermaphrodite) reproductive organs. In some rather more

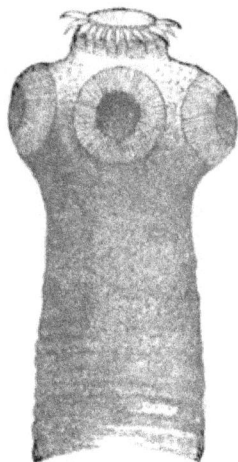  

Fig. 108. Fig. 109. Fig. 110.

Fig. 108.—Head of Tænia solium, showing suckers and circlet of hooks (from Leuckart).

Fig. 109.—Half-ripe and ripe joints of Tænia solium (from Leuckart).

Fig. 110.—Two proglottides of Tænia solium showing uterus (from Leuckart).

complex forms there are several sets of reproductive organs, but the external segmentation is incomplete. In all those that infest the intestine of man, there are numerous distinct joints, in each of which there is a single set of reproductive organs.

Ordinary Tape-worms may be said to occupy an intermediate position between segmented and unsegmented Worms, for though they have numerous similar joints, these are to a

14

certain extent independent of each other, and are readily got
rid of, without causing any apparent inconvenience to the
animal. A Tape-worm can therefore hardly be described as
a single, many-jointed organism, but must rather be termed a
collection or colony of organisms, only to a certain limited
extent mutually interdependent.

Life history of the Tænia solium.—The head of this worm is
furnished with a **circlet of hooks** and **four suckers,** by means
of which it fixes itself on to the mucous membrane of the
intestine of its host. As this is the only point of attachment,
the worm hangs freely in the intestinal canal. The terminal
joints, as they become ripe, drop off from the parent, and are
passed down the intestine to the exterior, sometimes one by
one, and sometimes several joined together. Each ripe joint
or proglottis is a sac, which has some power of muscular contraction, and which is distended with minute embryos enclosed
in firm egg shells. The proglottis, having been passed into
water or on to damp soil, soon ruptures, thus setting free its
brood of embryos enclosed in their shells. Should either the
unruptured proglottis or some of these eggs be swallowed by
a pig (the **intermediate host**), the resistent egg shells become
dissolved by the pig's digestive juices, and the embryos are set
free.

The embryos of the Tænia solium have been found in the Reindeer,
the Dog, and the Cat, but they only occur with frequency in the Pig and in
Man.

Each embryo has at its anterior end six hooks, by means of
which it attaches itself to the mucous membrane of the
intestine. It then at once begins to bore its way through the
wall; having accomplished this, it may fall into a blood-vessel
on the other side, and so be carried by the blood stream to the

liver, lungs, brain, eye, muscles, or rarely to other parts of the
body ; or it may drop into the body-cavity.

Having reached a resting place, the six-hooked embryo
begins to develop. It becomes a passive, vegetative, asexual
bladder-worm or proscolex, which after a few days is sufficiently large to be seen with the naked eye. Like any other
foreign body it causes a proliferation of cells around it, which

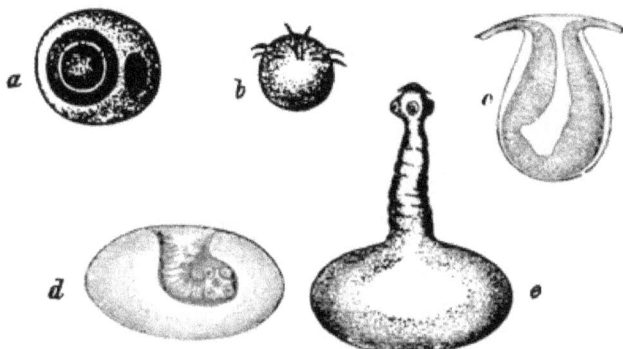

FIG. 111.—Stages in the development of Tænia solium to the Cysticercus stage
(after Leuckart and Claus). *a*, egg with embryo ; *b*, six-hooked free embryo ;
c, rudiment of the head as a hollow papilla on the wall of the vesicle ; *d*, bladder-
worm with retracted head ; *e*, the same with protruded head, magnified about
four times.

form a connective tissue sheath enclosing a smooth serous
cavity. As the embryo increases in size it becomes elongated,
its hooks drop off, and its central cells enlarge, become clear, and
then liquefy, forming a quantity of fluid, which fills the worm,
and produces the bladder-like appearance to which its name of
bladder-worm is due. In all the various Tape-worms, the name
of cysticercus or hydatid is given to the embryos at this stage,
the cysticercus of the Tænia solium being called the Cysticercus cellulosæ. If the resting place is in a muscle the

worms appear as small ovoid bodies, lying between the muscle-fibres.

At some part of the wall of the bladder there now appears a small thickening which soon grows inwards, forming a pouch in the bladder. The cuticle becomes invaginated at the same time, so that the appearance is that of a pushing in of the wall

FIG. 118.—Measly pork (natural size) (from Leuckart).

of the bladder at one part. This invaginated portion develops into the **head or scolex** of the mature animal.

At the same time that the **invagination** of the head is taking place, muscle is also developing in the walls of the head. At four points near the bottom of the invagination **four suckers** develop, and at the bottom of the cavity the **circlet of hooks** is produced. On the surface towards the bladder, a thin layer is seen to surround the head rudiment. This is called the receptacle. The rudiments of the water vascular system also develop about this time.

As soon as the head with its suckers and hooks, the rudiments of the water vascular system, and the muscle in the walls have been developed, the head becomes everted, so that the pouch now forms a projection on the top of the bladder; in other words, the pouch becomes turned inside out, with the result that the suckers and hooks, developed at the bottom of the pouch, now form projections on the outer surface of the head.

Having arrived at the stage of what appears now as a head with a bladder attached, development ceases for the time; that is to say, the animal remains quiescent. It only develops further after it has entered its final host. Pork containing cysticerci is known as measled. If this be eaten by another Pig or by a Man, either raw or in an imperfectly cooked condition, that is to say, if some or all of the cysticerci pass alive into the alimentary canal, the worms will develop into adults. As soon as a worm arrives in the stomach, the bladder and neck are dissolved by the gastric juice, the scolex being protected by calcareous concretions; the head, exhibiting active movements all the time, passes into the intestine, where it attaches itself, by means of its hooks and suckers, to the wall of the canal. Growth at once begins, as owing to its position it is richly nourished by its host, the head budding off asexually a chain of joints called the strobila. These joints are pushed farther and farther away from the head by the new joints, which are continually being formed. After some time the joints develop reproductive organs, which mature, the female organs giving rise to eggs, and the male ones to spermatozoa, which fertilise these eggs. When the eggs are ripe the mature joints drop off, and the life history of a new generation commences.

The life history of the Tape-worm is generally considered to form a good example of alternation of generations, the sexually mature proglottis giving rise to embryos which develop into scolices, the asexual animals from which the sexual proglottides are budded off. According to this view, the so-called adult animal, that is the animal which lives in the final host, is considered to be a colony of independent individuals, the scolex and its daughter proglottides. These latter only retain their connection with the scolex and with each other, whilst they are developing their sexual organs, and when a proglottis severs this connection, it is, so to speak, the first indication of its approaching death. It can, however, generally live for some time independently, and even sometimes increases considerably in size after it has dropped off. A strong argument in favour of this view is the fact, that although it is comparatively easy to get rid of all the proglottides budded off from the scolex, by giving drugs to the final host, it is by no means easy to get rid of the scolex itself, which, after losing the proglottides, at once begins to bud off new ones, being apparently quite uninjured by the violent removal of the ones previously formed. In the Tænia echinococcus several heads or scolices are budded off from the proscolex or bladder-worm, each head being capable of developing into an adult Tape-worm. Thus it is not the scolex which buds off sexual animals, but the asexual proscolex which buds off scolices, each of which gives rise to sexual proglottides.

The chief arguments against this view are—(1) that the segments are not really independent of each other as they have excretory canals (water vascular systems) and nervous systems in common ; (2) that it is only in the oldest segments that an independent existence can be carried on, and then only for a limited time, which in some forms is very short; (3) that in the portion of the animal immediately following the head there is at first no trace of distinct segmentation; (4) that the life history from the egg to the animal carrying sexually mature segments is the history of the life of a single individual, there being pauses, as it were, in the history but no breaks, and it is impossible to speak of an alternation of generations occurring in the life of a single individual. For an alternation to occur it is necessary that there should be a sexual individual alternating with an asexual individual.

To recapitulate—

1. The **six-hooked embryo** gives rise to the **proscolex or bladder-worm (cysticercus, hydatid).**

2. The **cysticercus** develops a head which becomes evaginated, and is then called the **scolex**.

3. The neck and bladder of the **cysticercus** are dissolved, and the scolex develops into an adult Tape-worm, which buds off a strobila of sexual proglottides, in which, when mature, eggs and spermatozoa are formed. These develop into embryos enclosed in their shells.

Of these phases, the first is passed partly in the body of the **final host**, as it is in his intestine that the egg develops into an embryo, enclosed in a shell. The shell is dissolved in the alimentary canal of the Pig, who acts as the **intermediate host**, · and it is in his muscles that the second stage occurs; whilst it is only after the animal has got into his **final host**, Man or another Pig, that the third phase of its life history is passed.

The above life history is true in the main, not only of the Tænia solium, but also of the Bothriocephalus latus, of the Tænia mediocanellata, and of the Tænia echinococcus.

The **Bothriocephalus latus** has no hooks and only two lateral suckers. Its final host is Man, its intermediate host being frequently the Pike. The embryo is ciliated, and swims freely. The cystic stage is absent, so that the scolex is really only a late stage of the embryo.

The **Tænia mediocanellata** has four suckers but no hooks. Man forms the final host, whilst the Ox is the intermediate host, giving lodgment to the cysticercus.

The **Tænia echinococcus** is a small Tape-worm, a quarter of an inch in length, with only four proglottides, and numerous though small hooks on its head. It occurs in the intestine of the Dog. Its bladder-worm (hydatid) is distinguished by the great thickness of the stratified cuticle, and by the fact that the bladder or hydatid cyst gives rise not to a single head but to a considerable number.

It usually occurs in the liver or lungs, but it may occur in any part of the bodies of domestic animals or of Man.

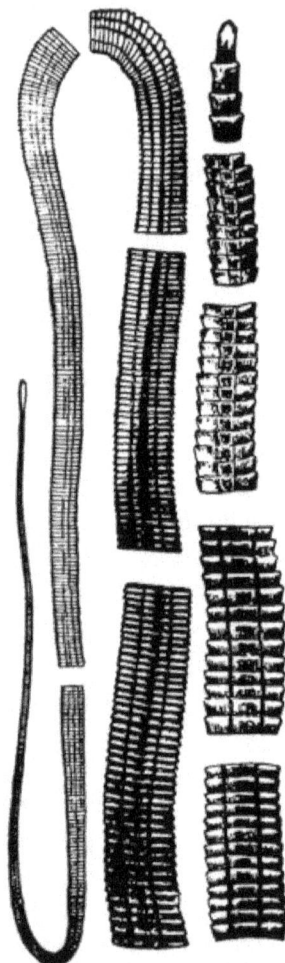

FIG. 113.—Bothriocephalus latus (after Leuckart).

When the embryo has reached the liver, it proceeds to

develop into a hydatid or cysticercus, losing its hooks, developing its bladder, and becoming invested by a connective tissue sheath formed by its host, exactly like the Tænia solium. The sheath grows with the growth of the hydatid. Generally,

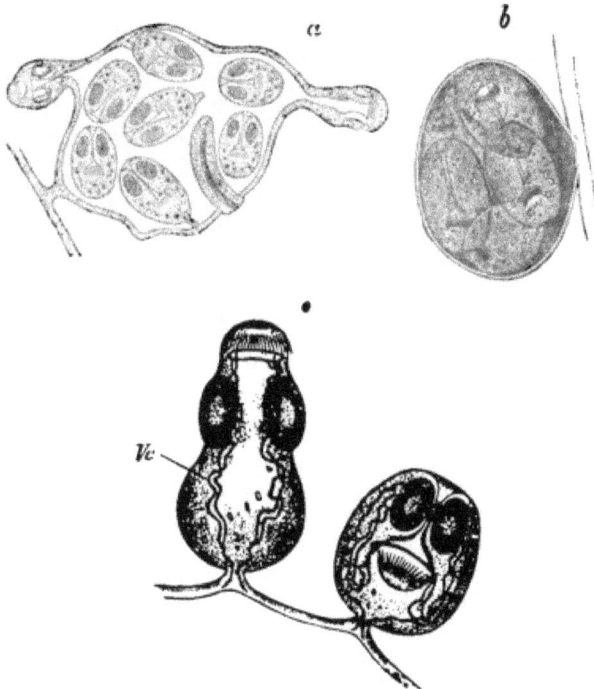

FIG. 114.—*a* and *b*, Brood capsules of Echinococcus with developing heads (after Leuckart and Wagener); *c*, heads of Echinococcus still connected with the brood capsule, one is evaginated; *Vc*, excretory canals (from Claus).

instead of a single bud, a number of buds are formed by the inner wall of the bladder. Each of these buds in turn forms a cystic body, which, for a time, remains attached to the wall by a pedicle. After it has become detached, it is called a daughter cyst, while the original bladder is termed the mother cyst. Each daughter cyst again may develop one or more granddaughter cysts in its interior. The original bladder thus becomes filled with a number of smaller bladders of various sizes ; sometimes thousands of them are present. Sooner or later the little buds cease to form detached vesicles and develop into small membranous sacs, the pedicles of which are persistent ; these are called brood capsules, because they give origin to a varying number of scolices or heads, each of which has its row of hooklets and its four suckers, and is capable, under favourable conditions, of growing into an adult Tænia echinococcus.

The hydatid cyst may grow to an enormous size, sometimes being as big as a child's head. These Echinococcus cysts occasionally remain sterile, when they are described as Acephalocysts.

Occasionally, though very rarely, a cyst is found which has given off numerous buds externally (Echinococcus multilocularis) ; these grow in all directions, and themselves give off new buds so that a mass of considerable size is produced, the original cyst appearing like a root, and the buds like so many branches. As these buds usually remain small, and are a good deal compressed by the tissues amongst which they force their way, the whole mass presents the appearance of a solid tumour, and it was at one time thought to be a peculiar species of cancer.

Anatomy.—The Tape-worm, which has been most thoroughly worked at, and which occurs most frequently in Europe, taken as a whole, is the Tænia mediocanellata, or, as it has also been called, the Tænia saginata. It is doubtful whether this Tapeworm or the Tænia solium is more common in England. One would have expected, that the pork Tape-worm would have been the commoner, from the fact that under-done pork is more frequently eaten than under-done beef.

The following description is that of Tænia mediocanellata,

but the differences between it and Tænia solium are not great.
An average specimen measures from fifteen to twenty feet in

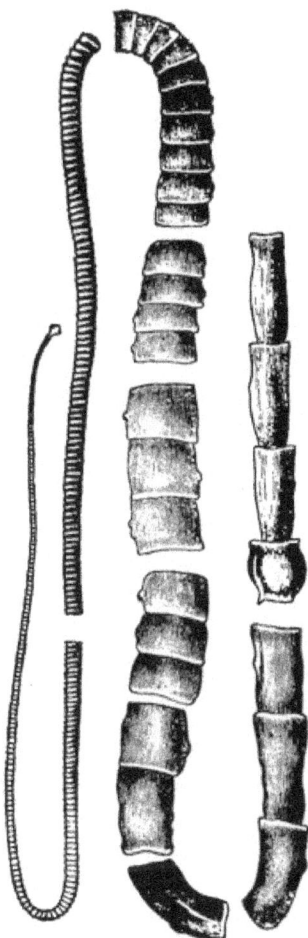

FIG. 115.—Tænia saginata (mediocanellata) (after Leuckart), natural size.

length, and consists of fourteen hundred segments; of these
more than two hundred and fifty will be found within a distance

of two inches from the head.   The segments are narrowest and
shortest near the head ; they widen and lengthen gradually until
near the middle of the chain they will be found to have a
width of about three-fifths of an inch and a length of about a
quarter of an inch.   Towards the terminal segments the width
decreases, being on an average about a quarter of an inch, but
the length continues to increase, until the terminal segment has
on the average a length of about four-fifths of an inch.   The
segments of this worm are also characterised by their firm
appearance and by their thickness.

The worm is sometimes called the unarmed Tape-worm, from
the fact that it has no hooks on the head.   The head is
spheroidal, about one-seventeenth of an inch in diameter, and
bears on its lateral surface four equidistant suckers by which
the animal attaches itself to the mucous membrane of the intestine of its host.   It has a flattened crown with a pit-like
hollow in the middle.   The suckers only project slightly, and
are surrounded by a pigmented border.

The **neck** (so called from its naked-eye appearance) has no
more right to its name, than has the scolex to the term head,
for it is only a succession of imperfectly developed segments.
It is generally about one-twenty-fifth of an inch long.   The
rings, which are at first very faintly marked and very narrow,
become more and more distinct and gradually larger the
farther they are removed from the head.   The sexual organs
become completely developed about the six-hundredth segment, while the embryos are only formed about three hundred
and fifty or four hundred segments farther on.   The simplicity
of the internal organisation corresponds with the simple appearance of the external structure of the majority of the segments,
it being only in the generative organs and in the excretory
apparatus that any degree of complexity occurs.

Beneath the external cuticle there is a groundwork of small cells, in which glandular cells are scattered here and there. Beneath this groundwork there is a delicate outer layer of longitudinal muscular fibres. Within this, again, there is a meshwork of connective tissue, in which all the organs are imbedded. At the outer part of this meshwork there are more longitudinal fibres, as well as an inner layer of circular muscle. Chiefly at the sides of the body, but to a small extent also in other parts, there are muscle fibres stretching

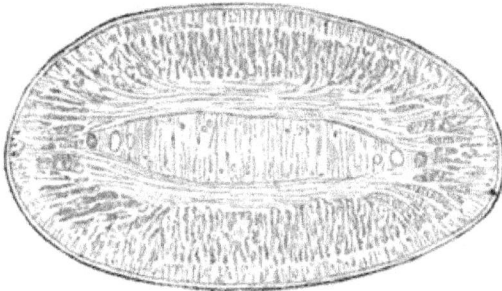

FIG. 116.—Transverse section of Joint of Tænia saginata (mediocanellata) (from Leuckart).

from one surface to the other. The changes in form which are seen to take place in the proglottis are due to these various layers of muscle, acting either independently or together. Thus the segment can shorten itself considerably, becoming at the same time broader and thicker; on the other hand, it is also capable of considerable elongation, when it becomes much thinner.

For a long time the existence of a nervous system was denied, but it has been proved to be present in the form of two lateral longitudinal cords, which run down externally to

the main trunks of the excretory system. These cords have
no sheath and hence are not very easy to see. In the head
they are somewhat swollen, and are connected together by a
transverse commissure, the swellings and the commissure

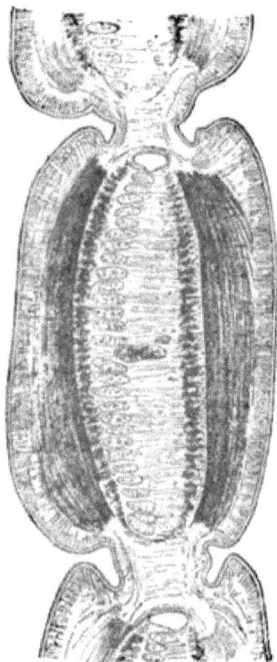

Fig. 117.—Longitudinal section of Tænia saginata (mediocanellata) (from
Leuckart).

being described as a cephalic ganglion. The nerve cords are
lined with delicate cells on their dorsal and ventral surfaces.
There are no distinct sense organs, but the skin, especially in
the head and suckers, is supposed to possess some sense of
touch. In one Tape-worm (Tænia perfoliata) nerve branches

have been seen to pass from the cephalic ganglion to the suckers.

The Alimentary Canal.—Nothing in the form of an alimentary canal is present. The animal has no need of a digestive system, as it takes in all nourishment through the walls of its body by osmosis, the substances being made ready for absorption by its host. It is probable, however, that some change is effected in the food by the cells in the body-wall during its passage.

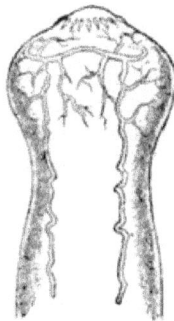

FIG. 118.—Head of Tænia serrata with its excretory vessels (from Leuckart).

The Blood Vascular System.—This is entirely wanting.

The Excretory System (water vascular system).—This is well developed, and is composed of a system of much-ramified canals distributed throughout the whole of the Tape-worm. It consists primarily of two longitudinal canals, one running along each side of the body. These are connected in the head and in each segment by transverse trunks. The longitudinal trunks appear straight, or wavy, according to the state of contraction or elongation of the animal's body. In many species there are four longitudinal trunks, but in both Tænia mediocanellata and Tænia solium there are only two. The transverse connecting trunks always occur at the posterior border of the joint (*see* fig. 119). The fluid in the longitudinal trunks can only flow in one direction—from before backwards,—its passage in the opposite direction being prevented by a valvular arrangement in the vessels.

Some of the fluid escapes when a segment drops off, as the vessels at the posterior end of the body are left open. At the same time, the sudden and complete emptying of the canals is apparently prevented by the curving forward of the short transverse trunk in this segment. Moreover, the ruptured ducts draw together when the division occurs. As soon as a segment drops off, the transverse trunk of the last segment but one begins to take on the same arrangement, so as to be ready for the next rupture of the ducts. The longitudinal canals are really efferent vessels for a system of very fine ducts, which ramify through the whole meshwork and receive long tubes, which begin as closed ciliated funnels. These canals contain a clear watery fluid without granules. It has been suggested that the longitudinal canals are the representatives of the cœlom or body-cavity, but this view has not yet been definitely proved.

Generative Apparatus.—Each proglottis is hermaphrodite, containing its own male and female generative organs, which are completely independent of the generative organs in any other segment. In the segments, close to the head, no sexual organs can be discovered ; and, in fact, it is only in about the four-hundredth joint, that the first traces of germ-producing organs are visible. The anterior and middle portions of the segment are seen to be granular, the granules being distributed pretty uniformly over the whole surface of these regions. These granules eventually become the male generative organs. In the posterior part of the segment, two plate-like masses of granules may be distinguished. Between them lies a third mass of somewhat different form. These develop into the two ovaries and the yolk gland. About two hundred segments further on, the male generative organs are found fully developed ; the female ones reach maturity a little later.

The **male apparatus** consists of a large number of pear-
shaped vesicles, called the **testes** or testicular vesicles ; these
are connected by their efferent ducts or vasa efferentia with a
common duct, the **vas deferens.** The coiled end of this duct
lies in a muscular pouch, called the **cirrus pouch** or sheath,
the duct itself being known as the **cirrus or penis.** The penis

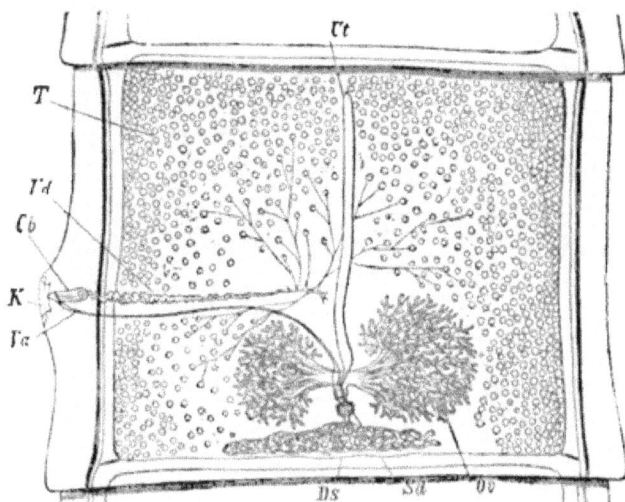

FIG. 119.—Proglottis of Tænia mediocanellata, with male and female organs (after
Sommer). *Ov,* ovary; *Ds,* yolk gland (vitellarium); *Sd,* shell gland; *Ut,*
uterus; *T,* testes; *Vd,* vas deferens; *Cb,* pouch of the cirrus; *K,* generative
cloaca; *Va,* vagina.

can be everted, and it is provided with spines. When everted
it protrudes, and then acts as a copulatory organ, entering the
vagina, generally of its own segment.

The **female apparatus** is very complicated. It consists of
a vagina, a shell gland, a yolk gland, a two-lobed ovary, a
receptaculum seminis, and a uterus.

The **vagina,** like the cirrus and vas deferens, is lined by
cuticle. It is a very fine tube, far too small to transmit the eggs

15

which, as before stated, are liberated by the bursting or dis-
solving away of the walls of the proglottis. At its inner end the
vagina loses its cuticular covering, and becomes dilated to
form a small swelling, to which the name **reoeptaoulum seminis**
has been given. Beyond this swelling, it is a rather dilated,
thin-walled tube, without any cuticular covering, and this short,
thin-walled portion, called by Leuckart the **fertilising canal**,
opens into the **shell gland** (Mehlis' body), which consists of
closely compressed nucleated glandular cells, provided with
thin ducts opening into the small internal cavity of the organ.
The radiate appearance of this organ is due to the regular
disposition of the cells and ducts. It is in this shell gland,
that the ovarian eggs acquire their outer envelope of yolk and
shell, the latter being probably the secretion of the glandular
cells. The shell gland has opening into it by a common
aperture, the fertilising canal and the common duct from the
two ovaries. It also communicates with the duct of the uterus
and that of the **yolk** or **albumen gland.** This organ, which
consists of a system of branched blind tubules opening into
efferent ducts, lies posterior to the shell gland, close to the
posterior limit of the segment. The cells lining the tubules
have well developed nuclei and nucleoli. The **ovaries** really
together form one organ, with two wing-like expansions, the
one nearer the vagina being the smaller of the two. In its
structure it resembles the yolk gland, except that its cells
(primitive ova) have a sharper contour, and in addition are
provided with a clear protoplasmic envelope; the primitive egg
has also a large nucleus (germinal vesicle), and the nucleolus
(germinal spot) is well marked.

The **uterus**, at this stage of development, consists of a
straight tube, which runs almost up to the anterior end of the
segment, and opens by a fine duct into the shell gland.

As before mentioned the vas deferens and the vagina both open into a common pouch or **genital cloaca**, which opens to the exterior by the **genital pore**. As soon as the testes have reached maturity, copulation takes place, and the spermatozoa from the testes are passed down through the vas deferens into the vagina, and so to the receptaculum seminis, where they remain until the female generative organs have reached maturity. As soon as the **ova** become fully developed, they

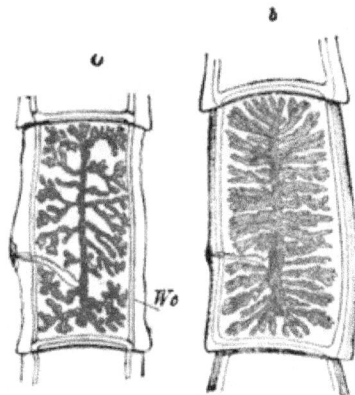

FIG. 120.—Ripe proglottides ready to separate—*a*, of Tænia solium ; *b*, of Tænia mediocanellata. *Wc*, water-vascular (excretory) canal (from Claus).

are fertilised by the spermatozoa, and then they pass into the uterus. As the uterus becomes distended by the rapidly growing eggs, it assumes its characteristic branched shape, and at the same time the testes, the ovaries, and all the rest of the generative apparatus more or less completely atrophy.

In the Bothriocephalus latus the genital pore lies on the ventral surface of the segment. In the Tænia solium it is placed alternately on the right and on the left side, so that it is impossible for neighbouring segments to fertilise each other.

It is not improbable that sometimes a segment fertilises the next but one to it.

Historical.—Some forms of Cestodes have been known from time immemorial. Aristotle speaks of bladder-worms in the tongue, and he and Hippocrates separated the Cestodes and other Flat worms from the Round worms. In the middle ages, the Tape-worm was known as Lumbricus latus, and all the varieties were considered to belong to the same species, until Felix Plater separated the Bothriocephalus latus from the other human Tape-worms. This Bothriocephalus was described by Bremser in 1819. Siebold in 1838 discovered the six-hooked embryo of the Tænia solium, and shortly after Steenstrup published his work on the alternation of generations. This at once suggested the explanation of the apparently contradictory statements made by various observers of the forms and life history of the Tape-worms. Küchenmeister in 1851 fed a dog with bladder-worms taken from a Rabbit, and a Cat with specimens from a Mouse, and he found that after an interval Tape-worms were present in the intestines of both these animals. Since 1851 numerous experimental observations have been made by Van Beneden, Leuckart, Mosler, St. Cyr, Cobbold, and many other observers, the results of which have furnished a very extensive list of hosts with their respective parasites.

CHAPTER XXII

THE **Nematoda** are **unsegmented** worms, with elongated cylindrical bodies, which taper at each end. With very few exceptions the sexes are separate, the male being generally smaller than the female. There is always a well-developed **cuticle**. A **nerve ring** surrounds the œsophagus, and from it there pass twelve nerves, six forwards and six backwards. The **alimentary** canal is usually well developed, and consists of a mouth, a muscular **œsophagus**, an **intestine**, a **rectum**, and an **anus**. There are two canals, one on each side of the body, between the dorsal and ventral groups of muscles. These may be **excretory** in function. They join together anteriorly, and open by a common anterior and ventral pore. The **male** usually has special copulatory spicules. The **reproductive** organs are long tubes, which open about the middle of the body in the females, and in the posterior region in the males by an aperture common to them and the rectum. The **spermatozoa** are small cells without tails. The **female sexual tube** is divided into the **vagina**, the **uterus**, the **oviduct**, and the **ovary**. The **male sexual tube** is divided into the **terminal testis**, the **vas deferens**, the **vesicula seminalis**, and the **ductus ejaculatorius**.

The **muscular system** is well developed, and consists of four distinct groups of muscles, two dorsal and two ventral, lying immediately beneath the epidermis. They are made up of band-like or fusiform cells, each of which consists of an internal granular portion and an external fibrillated layer.

The **body-cavity** is usually well developed, forming a space between the alimentary canal and the body-wall. **Cilia** are never present. There is no trace of a **blood vascular system.** In the free-swimming forms **eyes** are generally present, and also in these forms papillæ and tactile hairs occur in the neighbourhood of the mouth.

A few Nematodes, belonging to the family of the Anguillulidæ, are **free living**, spending their lives in damp earth or in water. Some are **parasitic** in or upon plants. Most, however, are parasitic upon animals.

In one form, that of the hermaphrodite nematode Rhabdonema nigrovenosum, the offspring of the parasitic worm attain sexual maturity in damp earth. They are then known as Rhabditis forms, and are peculiar in that the sexes are separate. Their offspring, however, grow up into hermaphrodite forms again.

The Gordiidæ, to which the Horse-hair Worm belongs, have in the adult condition no mouth, and a degenerate alimentary canal, although they swim freely in fresh water. Their larvæ are parasitic upon water insects, or upon animals which eat water insects, and have a well-developed mouth.

Development.—The development of the Nematodes is usually simple ; the egg after fertilisation develops into an embryo, usually while still within the egg shell. As a rule the eggs are laid before development has commenced, but in some forms (Filaria, Trichina) the animal is well developed before birth ; these forms are therefore viviparous.

The **parasitic Nematoda** usually inhabit the **digestive tract,** but in Vertebrates they also occur in the **lungs,** the **kidneys,** and **urinary bladder,** or in an encysted form in the **muscles** and **other organs.**

In some forms the larvæ develop directly into the adult. In many forms, however, they become encysted in the tissues of one animal, and develop into sexually mature adults in the digestive tract of another. Several species of Nematodes are human parasites; some of these are harmless, others are very dangerous.

The following forms frequently occur in Man : Ascaris lumbricoides, Oxyuris vermicularis, Trichocephalus dispar, Anchylostomum duodenalis, Trichina spiralis, Filaria medinensis, and Filaria sanguinis hominis.

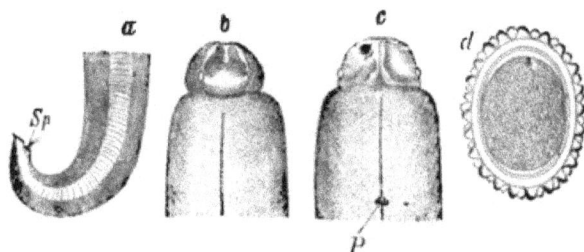

FIG. 121.—Ascaris lumbricoides (after Leuckart). *a*, Posterior end of a male with the two spicula (*Sp*) ; *b*, anterior end from the dorsal side, with the dorsal lip furnished with two papillæ ; *c*, the same from the ventral side with the two lateral lips and the excretory pore (*P*) ; *d*, egg with its external membrane.

The **Ascaris lumbricoides** (the common round worm or mawworm) is a cylindrical worm with pointed ends. It inhabits the intestine. The **female** varies from eight-and-a-half to sixteen inches in length. The **male** is considerably smaller ; its tail, which is bent to form a hook, bears two chitinous spicula. The **mouth** is surrounded by three muscular lips, armed with fine teeth. The female genital aperture is in front of the middle of the body on the ventral surface. The male genital aperture coincides with the anus. The female produces enormous numbers of eggs, which when mature are passed out with the fæces of the host ; they are provided with a double shell surrounded by an albuminous coating, and are very

FIG. 122.—Oxyuris vermicularis (after Leuckart). *a*, female ; *O*, mouth ; *A*, anus ;
V, genital opening. *b*, male with curved posterior end. *c*, the latter enlarged ;
Sp, apiculum. *d*, egg with enclosed embryo.

resistant both to cold and drying. The further history of the

animal is not known, but it is not improbable, that it has to
pass through an **intermediate host,** before it again visits man,
and develops into a sexually mature adult. The Ascaris lumbricoides is comparatively common in England, but is met with
far more frequently on the Continent. It often occurs **singly,**
but sometimes a very large number are present ; they are usually
eventually voided with the fæces, but are occasionally vomited
up. Unless present in very large numbers, this parasite does
no harm to his host, being rather a messmate than a parasite.

Oxyuris vermicularis, the thread-worm so frequently present
in the lower part of the intestine in children, is a small round
worm. The female is about two-fifths of an inch long, whilst
the male only measures about one-fifth of an inch. The female
has a pointed tail, whilst that of the male is blunt, and is
furnished with a single spicule. The egg is flat on one side
and round on the other, and is provided with a shell covered
by an albuminous coating. It must be swallowed by an
animal before it can develop into a sexually mature adult
worm. These parasites are frequently present in enormous
numbers, being passed in considerable quantities with the
fæces. They do no harm to their hosts, except by the violent
irritation they induce. The eggs are not destroyed by being
dried.

The **Trichocephalus dispar,** or whip-worm, is a small worm
inhabiting the lower part of the small intestine. Both male
and female measure about two inches in length. The anterior
part of the body is thread like ; the posterior part, which contains the genital organs, being much thicker. In the female
this posterior part is straight and cylindrical ; in the male it
is coiled into a spiral, and the tail is provided with a spicule.
The egg has a thick brown shell and a projection at each pole.
The larval stage is passed in water, and lasts for four or five

months. This parasite is not very often seen in England, but it is said to be very common in France. It is quite harmless.

Anchylostomum duodenalis, called also Dochmius duodenalis and Strongylus duodenalis, is a small worm inhabiting the upper part of the intestine. The female is cylindrical in form, a quarter to half an inch in length. The male is from one to two-fifths of an inch long. The head end is bent towards the

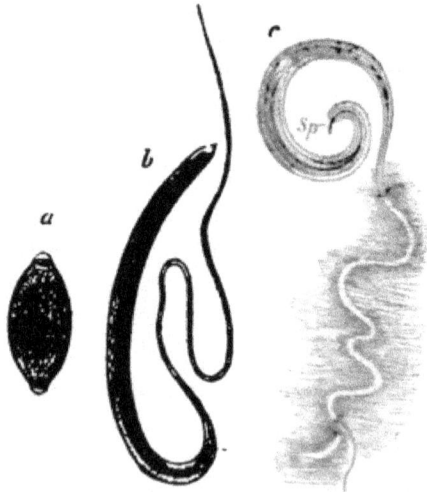

FIG. 123.—Trichocephalus dispar (after Leuckart). *a*, Egg ; *b*, female ; *c*, male with anterior part of body buried in mucous membrane ; *Sp*, spiculum.

dorsal surface, and has a strongly armed mouth provided with a horny oral capsule, armed with four curved and two straight teeth. The male genital opening is placed at the posterior end of the body, at the bottom of an umbrella or bell-shaped bursa ; the tail is provided with two thin spicules. The female becomes gradually thinner towards the posterior end, and terminates in a spine ; the genital aperture lies behind the middle of the body. The eggs are oval. The first stages of development take place in the human intestine ; the eggs then pass out with the fæces, and pass their next stage in water ; if they again find their way into the intestine of man, they rapidly develop, producing the mature worm, which feeds upon the

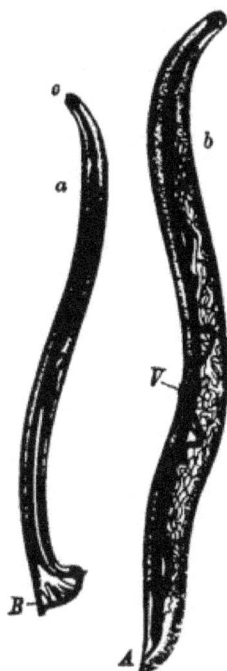

Fig. 124.—Dochmius duodenalis (after Leuckart). *a*, Male ; *O*, mouth ; *B*, bursa : , female ; *O*, mouth ; *A*, anus ; *V*, vulva.

blood of its host. This it obtains by gnawing a wound in the mucous membrane of the bowel. After the death of the host, the wound presents the appearance of a small swelling filled with blood, in the centre of which there is a small aperture which contained the head of the worm. Occasionally the swellings are larger, and contain coiled-up worms inside

them.   If present only in small numbers, no appreciable harm
results, but if there are large numbers of the parasites in the
bowel, serious and even fatal hæmorrhage may be produced.
The disease known as Egyptian chlorosis is produced by this
animal.

The **Trichina spiralis**, or flesh-worm of pork, occurs in
two forms, the immature form being found in muscles, etc., and
the mature in the intestine.   It is very minute, only about one-
ninth of an inch in length, and is very thin.   The posterior
portion in both sexes is straight ; the male is furnished at its
tail with two papillæ which are turned towards the ventral
surface, and which include between them four wart-like
nodules ; it has no spicule.   During copulation the rectum is
everted and protruded, so as to bring the aperture of the sexual
tube outside the animal's body.   The female genital pore
occurs at the junction of the anterior quarter with the posterior
three-quarters of the body.   The intestine is surrounded by
numerous cellular masses.   The eggs develop into embryos in
the uterus, and are born in the free state, the animal being
viviparous.   When set free they at once pass through the walls
of the intestine of their host, making their way either directly
into the peritoneal cavity or into the sub-peritoneal connective
tissues.   They may also pass into the blood and lymph streams,
and thus be conveyed to every organ of the body.   Those
that reach the muscles at once penetrate into the muscular
fibres, reducing the muscular tissue to mere organic *débris* ;
they take about fourteen days to develop into mature larvæ.
Each one then becomes surrounded with a cyst, formed partly
of the sarcolemma of the muscle, partly of a chitinous substance
secreted by the animal, and partly of connective tissue which
is formed owing to the irritation set up by the presence of the
parasite.   After a time calcareous salts are deposited in the

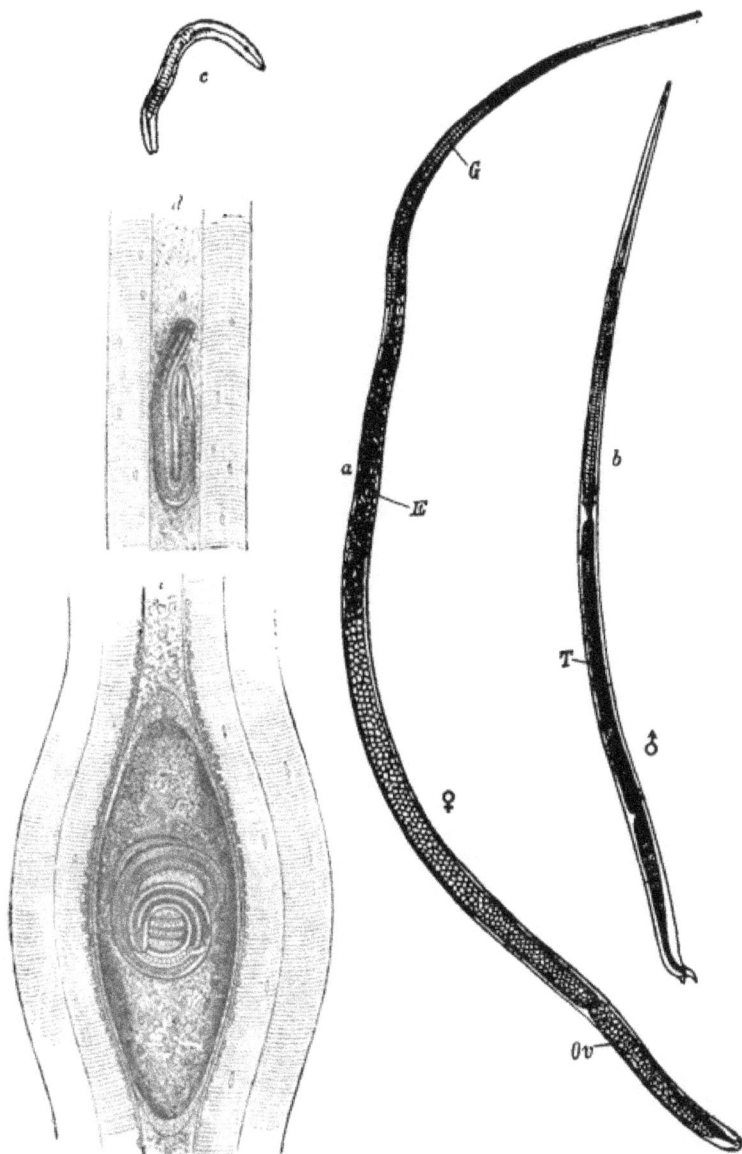

FIG. 125.—Trichina spiralis (from Claus). a, Mature female Trichina from the alimentary canal; G, genital opening; E, embryos; Ov, ovary. b, male; T, testis. c, embryo. d, embryo in a muscle fibre. e, the same developed into a coiled Muscle Trichina, and encysted.

cyst wall, so that the muscle appears to be dotted with innumerable minute white specks. If the Trichina dies, the contents of the cyst become calcified. The animal can, however, live for a very long time in this quiescent condition.

If uncooked pork, containing living encysted Trichinæ, be eaten by a human or other host, the cyst becomes dissolved in the stomach, after which the larvæ rapidly grow, reaching sexual maturity in about two and a half days. They then pair, and after the seventh day living embryos are born in large numbers, as many as a thousand being produced by a single female. The embryos then make their way through the walls of the intestine and into the muscles as above described.

Trichinæ have been met with in Men, Pigs, Rats, Cats, Mice, Foxes, Badgers, Hedgehogs, and a few other vertebrates. Man is usually attacked with trichinosis as the result of eating uncooked trichinous pork. If a large number of encysted Trichinæ have been swallowed, a fatal result usually ensues. The muscular Trichinæ are found most abundantly in the diaphragm, intercostal, cervical and laryngeal muscles, the muscles of the limbs being least affected.

The **Filaria (Dracunculus) medinensis**, or Guinea-worm, is a long thread-like worm from twenty-four to forty inches in length. The anterior end is rounded; the posterior terminates in a pointed tail curved over ventrally. The female only is known; it lives encysted under the skin of the leg or arm. The gravid uterus occupies the greater part of the body, and contains enormous numbers of ova and embryos; the latter have no envelope, their tails are pointed, and they possess a strong cuticle. The larval stage is passed in certain small Crustaceans (Cyclops) which live in drinking water. They are very common in Africa and Asia.

The **Filaria sanguinis hominis** inhabits the lymphatics

of Man, especially occurring in those of the scrotum and
lower limbs. They cause obstructions in the lymphatic
vessels, and give rise to the disease known as Elephan-

FIG. 126.—Filaria medinensis (after Bastian and Leuckart). *a*, anterior end seen
from the oral surface ; *O*, mouth ; *P*, papilla. *b*, pregnant female. *c*, embryos
strongly magnified.

tiasis. The lymphatics containing them may rupture, with
the result that lymph and parasites may escape into the urine
or into the abdominal cavity.

The adult female is about three inches long; it has a club-shaped head. The posterior part of the body contains a bifid uterus filled with ova and embryos, the latter being born alive. Within the uterus the embryo is provided with a membranous covering derived from the original envelope of the egg. As it grows it stretches this membrane, which, after the animal's birth, always remains attached to and investing it, projecting generally in the form of a lash from one end of the body. The intermediate host is the Mosquito, this animal taking in the embryos with the blood it sucks; the embryos pass a short time in the body of the Mosquito, and are then discharged into water. Here a further development takes place. If they then get into the body of their final host, they make their way into the lymphatics, and develop into sexually mature Filariæ.

CHAPTER XXIII.

THE LEECH (HIRUDO MEDICINALIS).

THE Leech belongs to the large group of animals included under the name Worms. Worms may be said to begin the series of cœlomate animals, being the lowest which possess a cœlom or body-cavity. The Leech belongs to the same large class as the common Earth-worm, although they differ from each other in many respects. Like the Earth-worm, the Leech is made up of a series of **segments**, each of which is, to a certain extent, a repetition of the one in front of it. Thus, in each segment is found a pair of nerve-ganglia upon the nerve-cords and a pair of excretory organs (nephridia). It will, however, be noticed that the external rings or annuli are very much more numerous than the number of segments; five annuli correspond to each segment or somite, there being about one hundred annuli, but only twenty-six segments in the body. The Medicinal leech is an elongated flattened worm. It varies from two to six inches in length, according as to whether it is stretched out or contracted. It lives in freshwater pools, in marshes, in sluggish streams, and either swims freely by vertical undulations of the body, or moves over the ground by attaching itself alternately by its anterior and posterior suckers. It lives on the blood of higher animals, such as Fishes and Frogs, to which it attaches itself for the purpose of sucking their blood. Its blood-sucking propensities are made use of in medicine for the purpose of abstracting blood from the human subject.

In shape, the Leech is either flattened or cylindrical, according as to whether it is contracted or not; it is widest just behind the middle, and, when flattened, is oval in transverse section, being more convex dorsally than ventrally. The skin of the segments, 9 to 11, is covered with a slimy secretion. This marks the position of the **olitellum** or saddle, the secretion of which makes the cocoon for the eggs.

The dorsal surface is darker than the ventral, and is usually marked on either side by three pale longitudinal stripes, the two outer of which, on each side, are marked by dark dots. The dots which occur on every fifth annulus are larger and darker than the rest, especially in the middle stripe on either side. That in which the dots are darker is the hindmost annulus of each segment of the body. The most anterior of the **five annuli**, which make up a **segment** or somite, has on it a transverse ring of small white dots. The ventral surface presents an irregularly mottled appearance.

There are **two suckers**, an anterior and a posterior one. The anterior one is placed on the ventral surface of the anterior end of the body, and is depressed in the centre to form the entrance to the **mouth**. The anterior edge of the sucker is frequently folded down over the mouth like a kind of lip. The posterior sucker is circular, larger than the anterior one, and is placed at the posterior end of the body, from which it is separated by a slight constriction. The muscle of the body-wall of the Leech is arranged as an outer circular layer, a middle oblique layer, and an inner longitudinal layer. There are, in addition, bundles of fibres crossing the body from the dorsal to the ventral surface (see fig. 128, p. 246).

Alimentary canal.—The **mouth** lying at the bottom of the depression in the anterior sucker is provided with **three jaws,** each of which consists of a muscular cushion covered on the surface by a horny cuticle. This cuticle is thickened at the free edge of the jaw, and is notched to form a large number of very fine teeth. When the animal has firmly fixed itself to its prey by its posterior sucker, it brings the anterior sucker down, and after taking a firm hold with it, it saws the skin with its jaws until a bleeding wound is produced. As all the jaws work, the wound is triradiate, each cut corresponding with the position of the jaw producing it. The blood is then sucked in through the mouth and through the very small aperture which leads from the mouth into the **pharynx.** The sucking action is produced by the pharynx whose walls are furnished with powerful muscles, the contractions of which serve to diminish the size of the pharyngeal cavity; the pharyngeal walls are further attached by the muscles to the body-walls, the contraction of these latter (radial muscles) serving to dilate the cavity. By the alternate contraction of the muscular walls and the radial muscles, blood

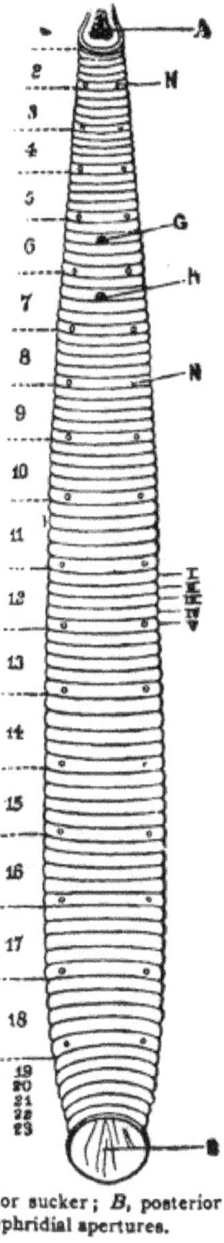

Fig. 127.—Hirudo medicinalis seen from the ventral surface (from Marshall after Bourne). The numbers 1 to 23 indicate the somites; the numbers *I* to *V* the five annuli of the twelfth somite; *A,* anterior sucker; *B,* posterior sucker; *G,* male aperture; *H,* female aperture; *N,* nephridial apertures.

is sucked in and forced on from the pharynx through a short
narrow tube, the **œsophagus**, into the next portion of the alimentary canal, the **crop**.   Upon the inner wall of the pharynx
lying amongst the muscles there are a number of large pear-
shaped cells, which secrete a fluid containing a ferment.   This
ferment prevents the blood which is sucked in from clotting.
These gland cells, by means of fine ducts opening close to the
jaws, pour their secretions upon the blood as it is taken in.

The **crop**, into which the œsophagus opens, is the most
conspicuous part of the alimentary canal.   It extends from the
fourth to the fourteenth segment, and consists of a straight, thin-
walled tube into which **eleven pouches**, whose walls are folded
internally, open upon either side.   These pouches are small in
the anterior region, but they gradually enlarge, until in the
posterior region they are of considerable size.   The hindmost
pair extend backwards for some distance on either side of the
intestine.

Just behind the crop, and communicating with it, there is
a small bilobed sac, which opens posteriorly into a narrow
straight tube.   The bilobed sac is **the stomach**, and the
straight tube **the intestine**.   The intestine is shut off from the
stomach by a ring of muscle forming a sphincter.   Considering
the large size of the crop and the large quantity of blood that the
animal can take in at one time, it is at first sight surprising that
the stomach should be so small, for it is probable that digestion
only takes place in the stomach, and that all the digested
material is absorbed through its walls.   But when we consider
that the Leech takes several months to digest a single meal,
it is evident that a small digesting and absorbing surface is
sufficient for its needs.

If a Leech is wanted to abstract more than one crop full of blood, it is
necessary to squeeze it from behind forwards so as to empty the pouches

into the crop; it will then vomit its sanguinary meal, and can, after an interval for rest, be again made to feed.

The intestine is a straight narrow tube, which runs from the stomach to an opening, the anus, which is a small aperture occurring on the dorsal surface of the animal just in front of the posterior sucker. In transverse sections of the Leech the alimentary canal can be seen to be made up of an outer connective-tissue layer containing a few muscular fibres, and an inner layer one cell thick, composed of short columnar cells.

The Renal Excretory system.—The organs, which correspond to kidneys in the higher forms and whose function is to excrete nitrogenous waste material, are the **nephridia** or segmental organs. Each segment from the second to the eighteenth contains a pair of nephridia, which lie at the sides of, and below the alimentary canal, and open to the exterior by minute apertures, a pair of which occur in the hindmost annulus of each segment on the ventral surface of the animal. Most of the nephridia commence as **cauliflower-like masses,** composed of spongy ciliated cells; these masses probably correspond to the ciliated funnels possessed by the nephridia of Worms that have a well-developed body-cavity. The mass of cells lies in a small sinus, the perinephrostomial sinus, and is continued into a rod, consisting also of a spongy mass of cells traversed in an irregular manner by very small ducts. In the segments which contain testes, this part of the nephridium lies in close relationship with the testis, and hence is called the **testis lobe.** The testis lobe is continuous with the **main lobe** of the nephridium, which consists of large granular cells and contains numerous fine-branched ducts. The main lobe is continuous with the **apical lobe,** which consists of cells the interiors of which have been hollowed out to form larger ducts; this lobe terminates the organ. A fine duct, consisting of cylindrical cells placed end

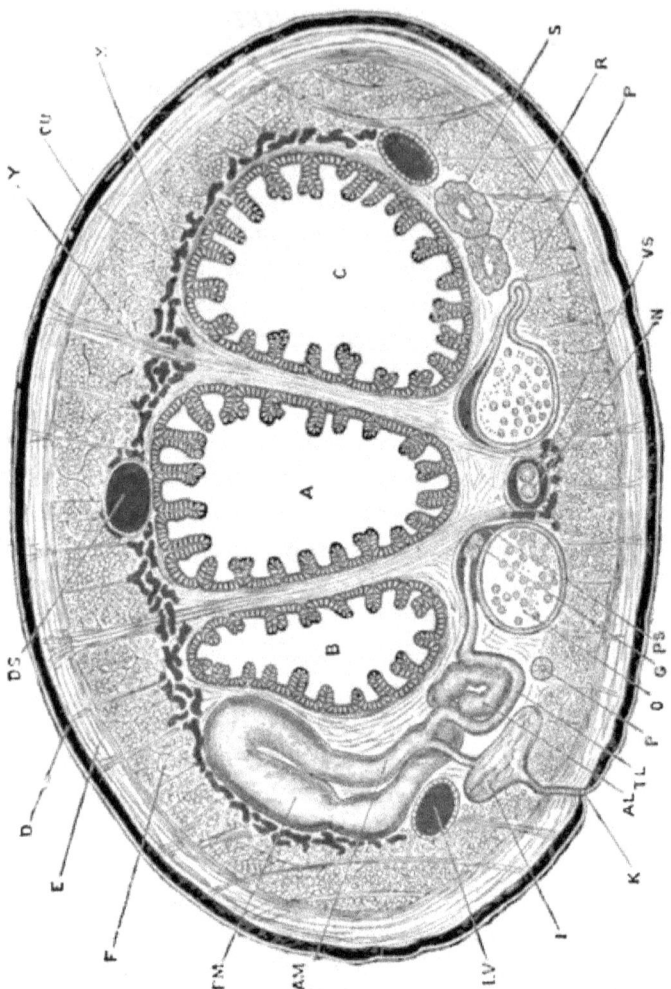

FIG. 128.—Hirudo medicinalis (from Marshall and Hurst). Diagrammatic transverse section through middle of body. *A*, crop ; *AL*, apical lobe of nephridium ; *AM*, anterior limb of main lobe of nephridium ; *B*, right and *C* left diverticula of crop ; *CU*, cuticle ; *D*, epidermis : *DS*, dorsal sinus ; *E*, layer of circular and oblique muscles ; *F*, longitudinal muscle layer ; *G*, cauliflower head of nephridium ; *I*, vesicle of nephridium ; *K*, external aperture of nephridium ; *LV*, lateral vessel ; *N*, nerve cord ; *O*, testis with developing spermatozoa ; *P*, vas deferens ; *PM*, posterior limb of main lobe of nephridium ; *PS*, perinephrostomial sinus ; *R*, section of nephridium ; *S*, intracellular ductules of nephridium ; *TL*, testis lobe of nephridium ; *VS*, ventral sinus ; *X*, botryoidal tissue ; *Y*, dorso-ventral muscles.

to end and hollowed out to form a tube, commences at the end of the apical lobe, and passes in a somewhat complicated manner through the organ, finally leaving it just at the junction of the main lobe with the testis lobe. It then runs as a short narrow tube called the **vesicle duct** to the **vesicle**, which is a dilated sac with muscular walls, and from which there runs a short tube which opens to the exterior. A connective-tissue sheath containing pigment invests the whole nephridium, and the organ is abundantly supplied with blood by branches from the lateral vessel.

The nephridia in segments two to seven, in which there are no testes, have very small testis lobes and no cauliflower-like ends.

The Vascular system.—This consists of the following parts:—

1. **Two lateral longitudinal blood vessels**, with well-developed muscular walls, which give off a large number of branches to the skin, to the alimentary canal, and to the excretory and reproductive organs.

2. **Two blood sinuses—the dorsal**, which runs along the dorsal surface of the whole length of the alimentary canal ; and **the ventral**, which runs beneath the alimentary canal. The walls of these sinuses are thin, and possess no muscle. Within the ventral sinus lies the nerve-cord.

The lateral vessels and the dorsal and ventral sinuses communicate with each other freely through capillary systems in the skin and in the various organs. The ventral sinus is in direct communication with the perinephrostomial sinuses. The lateral vessels communicate with each other by well-developed transverse vessels which pass across the body beneath the ventral sinuses. The dorsal and ventral sinuses communicate with each other by numerous fine vessels which pass posteriorly between the intestine and the crop.

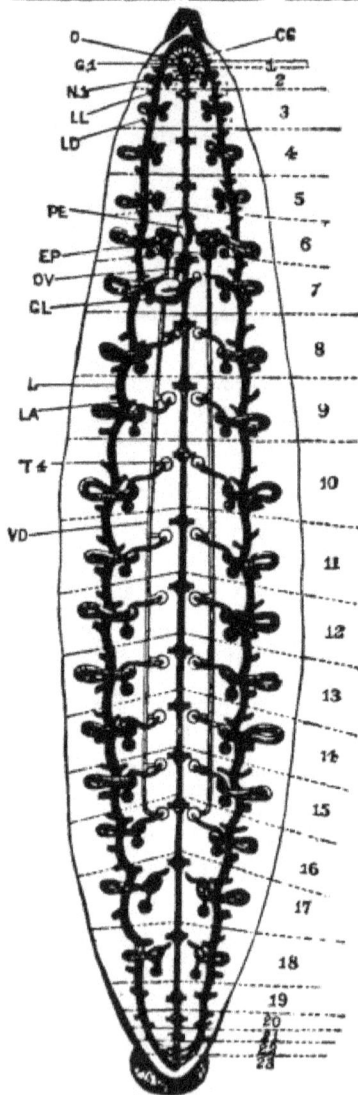

In addition to the capillaries in the skin and in the different organs there is a network of irregular channels, the **botryoidal tissue**, which occurs just below the longitudinal muscular layer of the animal, and is especially well developed around the crop.

The sinuses are really remains of the body-cavity, which, owing to the large development of connective and other tissue in the animal, is exceedingly reduced in size.

The blood is a red fluid in which float colourless amœboid corpuscles.

FIG. 129.—Hirudo medicinalis. Diagrammatic view of renal, nervous and reproductive systems after removal of alimentary canal (from Marshall after Bourne). The somites are numbered 1 to 23, and their boundaries indicated by dotted lines. *CG,* supra-œsophageal ganglion; *EP,* left epididymis; *G1,* the first of the twenty-three post oral ganglia; *GL,* glandular enlargement of the oviducts; *L,* lateral vessel; *LA,* latero-ventral branch of lateral vessel; *LD,* latero-dorsal branch; *LL,* latero-lateral branch; *N1,* the first of the seventeen nephridia of the left side; *O,* nerve collar; *OV,* ovisac containing the left ovary; *PE,* penis; *T4,* the third testis of the left side; *VD,* vas deferens.

Reproductive organs.—Both male and female organs occur in the same animal—that is, the Leech is **hermaphrodite**; but the egg cells are never fertilised by spermatozoa from the same animal, but always from another.

The male organs consist of nine pairs of **testes** lying in segments eight to sixteen. They are spherical sacs placed at the sides of the ventral sinus, towards the anterior portion of their respective segments. If the contents of a testis be examined with the microscope, it will be seen that they consist of spermatozoa in various stages of development. There are two thick-walled canals, the right and left **vasa deferentia,** with which each testis is connected by a short duct, the **vas efferens.** Each vas deferens runs longitudinally along the ventral body-wall, ending posteriorly blindly in the sixteenth segment, and anteriorly in the sixth segment in a coiled tubular body, the **epididymis,** which secretes a viscid fluid for the purpose of gluing the spermatozoa together into packets, called spermatophores. Each epididymis leads by a short duct on its inner side to the base of the **penis,** which is a pear-shaped tubular body, also lying in the sixth segment. The penis can be protruded some distance out of the male

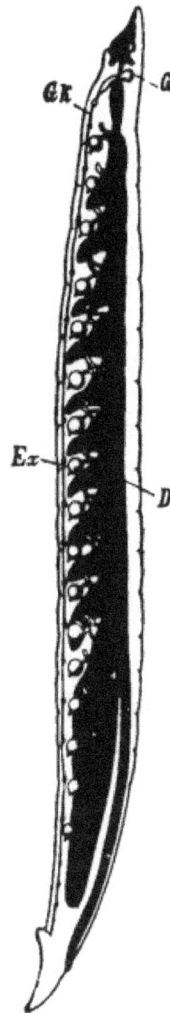

FIG. 130.—Longitudinal section through the Medicinal leech (after Leuckart). *D*, intestinal canal; *G*, cerebral ganglion; *Gk*, ganglionic chain; *Ex*, excretory canals or segmental organs (nephridia).

genital aperture which lies in the anterior part of this
segment.

The female organs do not present the same segmental repetition as the testes, the whole apparatus being contained in
the seventh segment.   They consist of—

1. Two minute **ovaries**, enclosed in the ovisacs, which lie
close to the middle line on the ventral surface.

2. Two **oviducts**, narrow tubes, which lead from their respective ovisacs, and unite with each other to form a single
much convoluted duct with glandular walls.

3. The **vagina**, connected by this single united duct with
the oviducts, and opening to the exterior at the female genital
pore in the anterior part of the seventh segment.   It is a thick-
walled, muscular tube.

The favourite breeding-time of the Leech is in spring.   Two
Leeches unite head to tail so that the penis of each enters the
vagina of the other.

The **Nervous system** consists of two ganglia (**supra-oesopha-
geal**) lying close to each other on the dorsal wall of the
anterior end of the pharynx.   From these, two nerve-cords
run round the pharynx (**nerve-collar**) and down the whole
length of the animal.   They are so close together, that they
appear to form only one cord, and each pair of ganglia appears
to unite to form a single ganglion.   The first pair of ganglia
(the sub-œsophageal) are the largest, and give off five pairs of
nerves; the next twenty-one pairs are small and give off two
pairs of nerves each, whilst the last or twenty-third pair are
also large and supply the posterior sucker.   The jaws, eyes,
and surrounding parts are supplied by branches from the supra-
œsophageal ganglia.   From the beginning of the ventral chain
special nerves are given off to the alimentary canal to form a
visceral nervous system.

Sense organs.—In addition to the dots already mentioned, which are supposed to have some sensory function, there are on the first eight rings ten eye spots. These consist of cup-shaped depressions in the epidermis, the base of each cup being surrounded by connective tissue containing black pigment. The central cells of the cup are in direct connection with nerve fibrils, and are surrounded by large clear cells, which may possibly have some function in condensing the rays of light upon the sensory cells.

In a microscopical section of the skin of the Leech, the epidermis can be seen to consist of a single layer of columnar cells, on the outer surface of which is a structureless cuticle, and amongst which numerous unicellular gland cells occur. Below this is the dermis, which consists of a modified connective tissue, containing numerous pigmented fibres, some muscle fibres irregularly arranged, and an abundant plexus of blood vessels. Some of these blood vessels penetrate between the inner ends of the epidermal cells. By this means the blood is brought into close relationship with the oxygen dissolved in the water. Thus the aeration of the blood is provided for, and the necessity for special organs of respiration is avoided.

CHAPTER XXIV.

THE DOG-FISH (SCYLLIUM CANICULA).

THE Dog-fish may be taken as a convenient type to illustrate some of the more important characters of the lower vertebrates. It is **bilaterally symmetrical**. It is provided with four longitudinal fins (two dorsal, a caudal, and a ventral), and also an anterior and a posterior pair of lateral fins; these represent the anterior and posterior limbs of the higher animals.

It has the following apertures: a mouth, a cloacal aperture, paired nostrils, spiracles, and, along the sides of the neck, five pairs of gill clefts. At the sides of the cloaca, a pair of pores occur, which in the adult open into the body-cavity.

The **skeleton** is entirely cartilaginous, and consists of a skull; a well-developed vertebral column; short ribs in the anterior part of the body; a **visceral skeleton** encircling the anterior part of the alimentary canal, and bearing gill rays; and anterior and posterior pelvic girdles, which are only slightly developed, and articulating with which are the lateral fins. The transversely elongated mouth leads into a wide **pharynx**, from which a short **œsophagus** leads into the U-shaped **stomach**. From the stomach a very short **small intestine** leads into a comparatively short **large intestine**, provided with a spiral valve. At its hinder end it narrows to form the **rectum**, which opens into the **cloaca**.

The **liver** is a large bilobed organ, from which a duct passes to open into the small intestine.

The **pancreas** is a small organ lying between the stomach and the intestine, with a duct opening into the latter.

The **spleen** is a large brownish-red body attached to the stomach.

The **heart** is branchial, lying in a pericardial cavity just below the pharynx. It consists of a sinus venosus, an auricle, a ventricle, and a truncus arteriosus. From the latter the efferent gill vessels pass; these, after breaking up in the gills, re-collect to form the dorsal aorta, which therefore contains aerated blood, which, after passing through the body, is returned to the heart by venous sinuses—namely, the anterior cardinal sinuses, the inferior jugular sinuses, the posterior cardinal sinuses, and the hepatic sinuses. The posterior cardinal, anterior cardinal, and inferior jugular sinuses pour their blood into a small pouch on either side, the Cuvierian sinus; from the Cuvierian sinuses it passes into the sinus venosus. The blood from the stomach and intestines passes through the portal vein to the liver; from this organ it is returned to the sinus venosus through the hepatic sinuses. The blood is returned from the tail by the caudal vein. This vessel divides to form two renal portal veins, which break up into capillaries in the kidneys. From the kidneys the blood passes by means of the posterior cardinal sinuses to the Cuvierian sinuses, and so to the heart. The blood is red, and contains both red corpuscles and white corpulscles.

The **kidneys** are two elongated bodies placed in the posterior part of the body, on either side of the middle line, behind the thick peritoneum. Their ducts (ureters) open into sinuses which open into the cloaca.

Reproductive organs.—The **ovary** is single. There are

two oviducts united together by their inner ends, and communicating by a single opening with the abdominal cavity. The opening occurs just at the junction of the two tubes, and is placed immediately in front of the liver. At their posterior ends the two oviducts unite, and open into the cloaca. Upon each oviduct is a large oviducal gland, which secretes the horny capsules of the egg.

The male has a **pair of testes** placed in the abdominal cavity. Each testis opens by numerous fine ducts (*vasa efferentia*) into an organ which lies in direct line with, but in front of, the kidney of its side. This organ is the **mesonephros**, which in the Dog-fish functions as the epididymis. The duct from the mesonephros is the **vas deferens**, and it dilates to form both a **vesicula seminalis** and a **sperm sac** at its posterior end. The two sperm sacs unite together to form a sinus, the urinogenital sinus, into which the ureters open ; this sinus then opens into the cloaca.

The Nervous system.—The brain can readily be exposed by slicing off the roof of the skull ; it can then be seen to consist of **olfactory lobes**, small **cerebral hemispheres**, a **thalamencephalon**, two **optic lobes**, a **cerebellum**, and a **medulla oblongata**. This is continued as the **spinal cord**.

The cerebral nerves can be very readily studied in this animal. The first pair go to the olfactory organs, the second to the eyes ; the third, fourth, and sixth pierce the wall of the orbit and pass to eye muscles ; the fifth and seventh each give off a branch (ophthalmic divisions) which pierce the orbit and run forwards on its inner wall ; the fifth further divides in the orbit into two divisions, which fork over the angle of the mouth ; the seventh also divides into two more branches, one of which goes forward to the palate, while the other forks over the spiracle ; the eighth goes to the auditory organs ;

the ninth forks over the first gill slit, and the tenth gives off branches which fork over the remaining gill slits, and is then itself continued on to supply the heart and stomach.

The spinal nerves have two roots, a single dorsal one and a multiple ventral one.

The eyes, olfactory organs, and auditory organs are well developed, but no external auditory apertures occur.

CHAPTER XXV.

GENERAL REVIEW OF THE MAMMALIA.

THE main characteristic of the Mammalia is that the young are nourished, for a variable period after birth, by the mother's milk.

Mammals are divided into three classes : the Monotremata, the Marsupialia, and the Placentalia.

The Monotremata include the Duckmole (Ornithorhynchus) inhabiting Australia and Tasmania ; the Echidna found in Australia, Tasmania, and New Guinea, and the Pro-echidna living in New Guinea. In these animals the skin is covered with soft fur, or hairs and spines, the skull is smooth, the sutures being obliterated, the surface of the brain is not convoluted, the alimentary canal ends in a cloaca, the ureters open not into the bladder, but into the urethra, and the testes in the male do not come down into a scrotum, but remain in the abdomen. In the female the left ovary is larger than the right, and the young are hatched outside the animal—that is, the Monotremata lay eggs.

The Marsupialia include the Kangaroo, Opossum, etc. In these animals the skeleton presents several peculiarities ; there are two bones in front of the symphysis pubis called the epipubic or marsupial bones. The scrotal sac lies in front of the penis, and the period during which the embryo remains within the uterus of the mother is short (five weeks in the

FIG. 131.—Skeleton of Greenland whale (Balæna mysticetus) (after Eschricht and Reinhardt). *Ocs*, occipital bone; *Co*, occipital condyle; *Sq*, squamosal; *Pa*, parietal; *Fr*, frontal; *Jmx*, praemaxilla; *Mx*, maxilla; *J*, jugal; *L*, lachrymal; *St*, sternum (only connected with first rib); *Sc*, scapula; *H*, humerus; *B*, rudiment of pelvis; *F*, of femur and *T*, of tibia.

17

Kangaroo). The young are born prematurely, and are then carried in the external pouch or marsupium for some time, being nourished by the mother.

The **Placentalia** are characterised by the close connection by means of a placenta between the uterus of the mother and the embryo during gestation. The blood of the mother and the blood of the fœtus come into close relationship with each

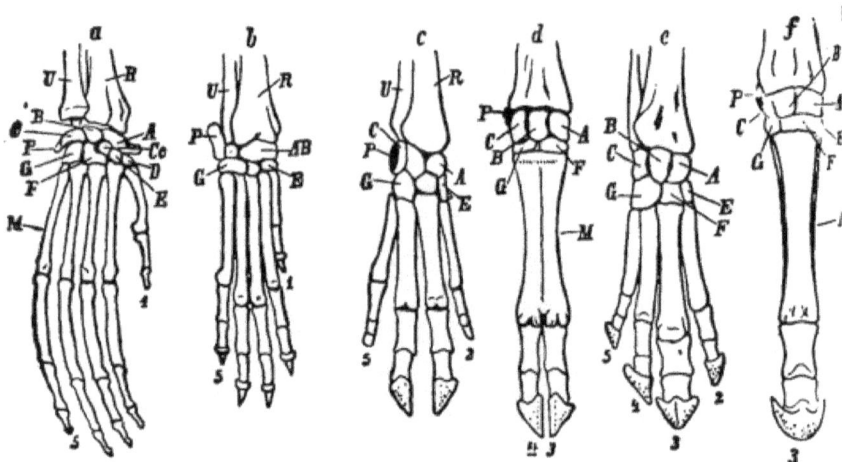

FIG. 132.—Skeleton of a hand of—*a*, Orang; *b*, Dog; *c*, Pig; *d*, Ox; *e*, Tapir; *f*, Horse (after Gegenbauer and Claus). *R*, radius; *U*, ulna; *A*, scaphoid; *B*, semi-lunar; *C*, cuneiform; *D*, trapezium; *E*, trapezoid; *F*, osmagnum; *G*, unciform; *P*, pisiform; *Cc*, centrale; *M*, metacarpus.

other, the fœtus obtaining all its nourishment, including oxygen, from the blood of the mother.

All Mammals are quadrupeds, with the exception of the Sirenia (Sea-cows) and Cetacea (Whales, Dolphins), in which the hind limbs have either disappeared or are rudimentary, but Man is the only Mammal, which habitually walks erect.

The external appearance differs considerably in different orders; whilst some Mammals are aquatic, and have a fish-like appearance (Whales, Porpoises, Dolphins, Sea-cows); others, such as the Ornithorhynchus, have bodies adapted for both land and water, and again others are only fitted for a purely terrestrial life.

The Cetacea are peculiar in possessing no external ears. The extremities of Mammals present several variations, as to their length, as to the amount of fission with each other of different bones, and as to the complete disappearance of these bones. For instance, the **hind limbs** are only rudimentary in Cetacea.

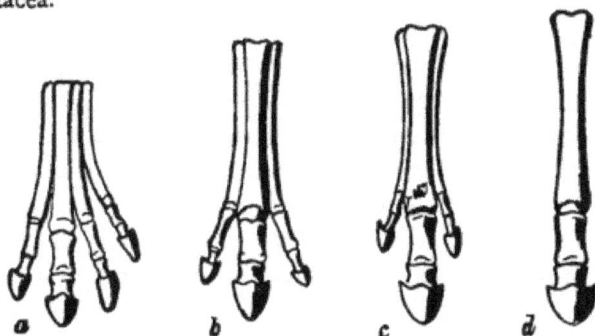

FIG. 133.—Pedal skeleton of different genera of Equidæ (after Marsh). *a, b,* and *c* fossil forms; *a,* foot of Orohippus (Eocene); *b,* foot of Anchitherium (Lower Miocene); *c,* foot of Hipparion (Pliocene); *d,* foot of the recent genus Equus.

In the large majority of cases the modifications occur chiefly in the foot and hand. In the family of the Equidæ, to which the Horse belongs, the three-jointed middle digit alone touches the ground, the other toes having either partially or entirely disappeared; but fossil forms are known in which there was not nearly so great a reduction of parts.

Some Mammals, such as the Bat, are enabled to fly by the formation of a membrane, the patagium, which is an extension of the skin, and which stretches between the fore limbs and the body, and between the very long fingers, thus forming wings.

In the swimming Mammals the fore limbs are modified to form flipper-like organs, which function as fins. In all Mammals with the exception of the Monotremata, the coracoid bone is only represented by a process from the scapula (coracoid process).

Skin.—Hairs are never entirely absent, although they are scanty in some forms, and in others only present in early life.

In the Armadillo alone, amongst Mammals, there is a bony exoskeleton.

Dentition.—Mammals differ from each other considerably in the number and character of their teeth. These generally either last throughout life, or are only once changed. The following is the typical mammalian dentition: three incisors, one canine, four præmolars, and three molars on each side of either jaw. This is represented as a formula where $i =$ incisors, $c =$ canines, $pm =$ præmolars, and $m =$ molars, the numerators representing the condition in the two sides of the upper, and the denominators of the lower jaw. Thus the typical dentition is

$$i. \frac{3-3}{3-3}, \quad c. \frac{1-1}{1-1}, \quad pm. \frac{4-4}{4-4}, \quad m. \frac{3-3}{3-3} = \text{a total of } 44 \text{ teeth.}$$

The following are the formulæ in a few Mammals :—

OPOSSUM	$i. \frac{5-5}{4-4}$,	$c. \frac{1-1}{1-1}$,	$pm. \frac{3-3}{3-3}$,	$m. \frac{4-4}{4-4}$.
DOG	$i. \frac{3-3}{3-3}$,	$c. \frac{1-1}{1-1}$,	$pm. \frac{4-4}{4-4}$,	$m. \frac{2-2}{3-3}$;
PIG AND HORSE	$i. \frac{3-3}{3-3}$,	$c. \frac{1-1}{1-1}$,	$pm. \frac{4-4}{4-4}$,	$m. \frac{3-3}{3-3}$.
SHEEP	$i. \frac{0-0}{3-3}$,	$c. \frac{0-0}{1-1}$,	$pm. \frac{3-3}{3-3}$,	$m. \frac{3-3}{3-3}$,
CAT	$i. \frac{3-3}{3-3}$,	$c. \frac{1-1}{1-1}$,	$pm. \frac{3-3}{2-2}$,	$m. \frac{1-1}{1-1}$.
MAN	$i. \frac{2-2}{2-2}$,	$c. \frac{1-1}{1-1}$,	$pm. \frac{2-2}{2-2}$,	$m. \frac{3-3}{3-3}$.

The shape and character of the teeth vary according to the diet of the animal; thus the Dolphin has a large number of uniform small teeth suited to grasp a slippery body like a Fish. In the baleen Whales the teeth are shed before birth, their place being partly supplied by plates of whalebone (baleen) developed on the palate. The carnivorous Mammals, like the Dog, have long and sharp canines, suitable for tearing food. The insectivorous Mammals have incisors which meet very accurately, to enable the animal to secure small active prey. Herbivora, like the Sheep, have incisors suited for cropping grass. In Man, as in other omnivorous Mammals, the teeth are fitted for masticating different kinds of food.

The **tongue** is never absent, but it may be fused with the floor of the mouth, as in Whales. It may be specially adapted for seizing, as in Giraffes, or for capturing food, as in Ant-eaters. **Salivary glands** are present in all land forms, but are rudimentary or absent in the Whale. In Herbivora very large quantities of saliva are poured out for the purposes of moistening the food, and converting the starch into sugar. The sides of the **buccal cavity** are soft and fleshy, and are dilated in some Rodents and Apes into wide sacs, the so-called cheek pouches. The soft palate, separating the posterior part of the nasal cavities from the pharynx, is peculiar to Mammals. The **œsophagus** is usually long, and opens into the stomach. The **stomach** is usually, as in Man, a simple sac, but it is often subdivided by the enlargement of certain parts of it, and the partial separation of these parts from each other to form several chambers. Thus in Ruminants there are four divisions of the stomach, the paunch or rumen, the honeycomb bag or reticulum, the manyplies or psalterium, and the reed or abomasum. The food first passes into the paunch, which is characterised by having numerous villi on its surface; it is here softened,

and is later regurgitated into the mouth. It is then well
chewed (chewing the cud) and mixed with saliva. It is again
swallowed, and now passes directly through the reticulum into
the psalterium. The honeycomb bag, or reticulum, owes its
name to the honeycomb-like appearance of its internal walls.
The psalterium, or manyplies, functions as a filter; its walls

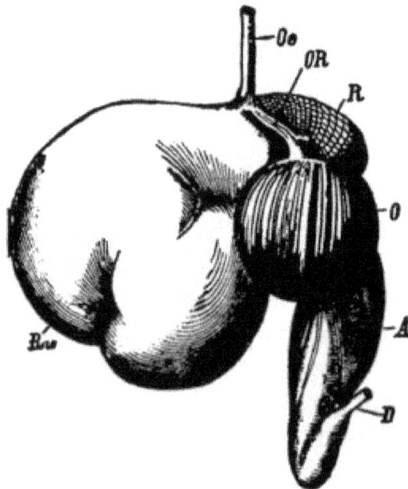

Fig. 134.—Stomach of a Calf. *Ru*, Paunch or rumen; *R*, reticulum; *O*, psalterium
or manyplies; *A*, abomasum or rennet stomach; *Oe*, end of œsophagus;
OR, œsophageal groove; *D*, beginning of intestine (from Claus).

are raised into numerous leaf-like folds. From the psalterium
the food passes directly into the reed or abomasum, which is
the true digestive stomach secreting the gastric juice. Between
the small and large intestines there is a valve and a pouch, the
cæcum, which is small in Man, but in Herbivora is very largely
developed. The small intestine is long in Herbivora, short
in Carnivora, and of medium length in Omnivora.

The **Respiratory organs** consist of paired lungs enclosed in

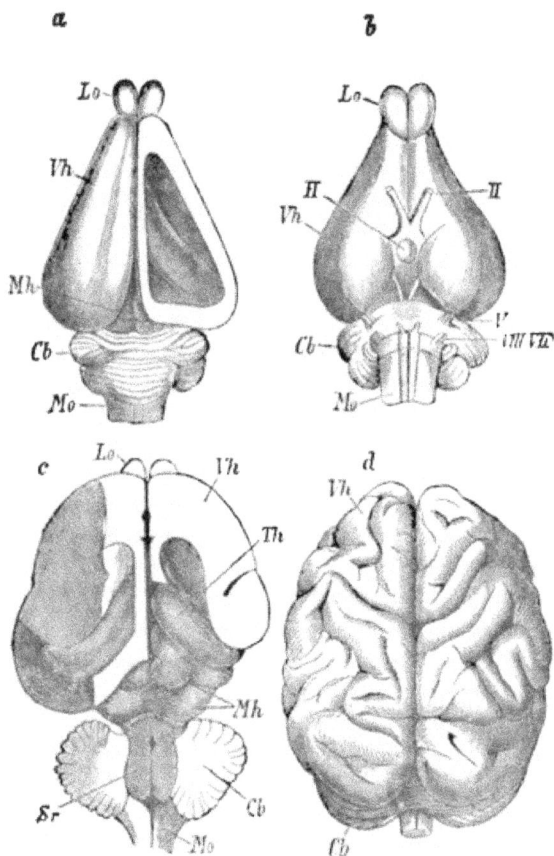

FIG. 135.—Brains of Mammals (*a*, *b*, and *c*, after Gegenbauer; and *d*, règne animal). *a*, of Rabbit, from above, with roof of right hemisphere removed so as to expose the lateral ventricle; *b*, the same, from below; *c*, brain of Cat, on the right side the lateral and posterior part of the hemisphere is removed and almost as much on the left side, the greater part of the hemispheres of the cerebellum have also been removed; *d*, brain of Orang. *Vh*, cerebral hemispheres; *Mh*, corpora quadrigemina; *Cb*, cerebellum; *Mo*, medulla oblongata; *Lo*, olfactory lobe; *II*, optic nerve; *VN*, trigeminal; *VII, VIII*, facial and auditory nerves; *H*, hypophysis cerebri; *Th*, optic thalamus; *Sr*, sinus rhomboidalis.

a thorax, which is separated from the abdominal cavity by a muscular diaphragm.

The **kidneys** occasionally consist of numerous lobes joined together at the pelvis, as in seals and dolphins, but usually they are compact bean-shaped organs.

Reproduction.—In Ungulates, or hoofed animals, the young are born in a very highly developed condition, being capable of running almost immediately after birth. In the Cetacea the young are also very highly developed when they are born. In Man and in many other Mammals, the young are for a considerable time dependent upon their parents for protection, being quite unable either to escape from, or to defend themselves against enemies.

The **oviducts** of the female, except in Monotremes, unite at their lower ends for a variable extent. The young live during gestation in the lower parts of the separate or united oviducts, these parts being known as the single or double **uterus.** The upper parts of the oviducts are known as the **Fallopian tubes.**

In different species there are differences in the extent to which the lower (uterine) portions of the oviducts are joined together—(1) They are quite separate as in Rodentia (Rabbits, Hares, Squirrels, Dormice, Rats, Mice, Guinea-pigs, Porcupines, etc.); (2) only the horns of the uterus are separate, the uterine portions of the oviducts being largely fused, as in Ungulata (Pigs, Hippopotamuses, Camels, Cattle, Deer, Tapirs, Rhinoceroses, Horses, and Elephants, etc.), Carnivora (Lions, Tigers, Foxes, Dogs, Bears, Otters, Seals, and Walruses), Cetacea (Whales, Dolphins, and Porpoises, etc.), and Insectivora (Hedgehogs, Moles, Shrews, etc.); or (3) there is only a single cavity with muscular walls, the Fallopian tubes being continued right down to the single uterus, as in Primates (Lemurs, Marmosets, American monkeys, Baboons, Anthropoid apes, and Man). There may be several pairs of **mammary glands** or only a single pair as in Man.

The cerebral hemispheres are always well developed in Mammals. In the lowest forms they are smooth on the surface; in the somewhat higher forms they become more complex, owing to the formation of ridges and depressions on the surface; and in the highest forms, as in Man, the ridges have become convolutions (gyri) separated from each other by deep grooves (sulci).

The **Pons varolii** is but little developed in the lower forms,

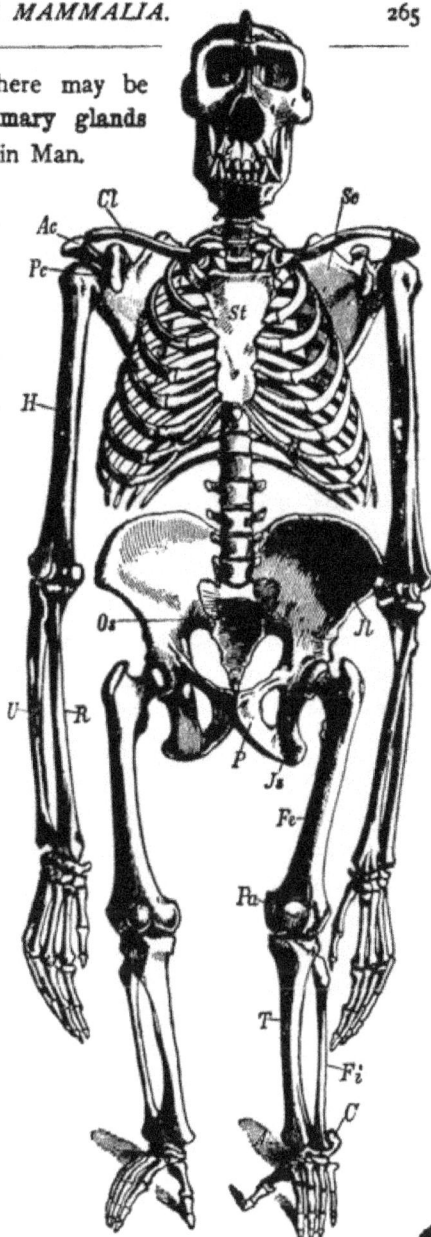

FIG. 136.—Skeleton of Gorilla (from Claus). *St*, sternum; *Sc*, scapula ; *Ac*, acromion; *Pc*, coracoid process; *Cl*, clavicle ; *H*, humerus ; *R*, radius ; *U*, ulna ; *Os*, sacrum ; *Jl*, ilium ; *Js*, ischium ; *P*, pubis ; *Fe*, femur ; *Pa*, patella ; *T*, tibia; *Fi*, fibula ; *Ca*, calcaneum ; *A*, astragalus.

whereas in the higher forms it is increased, into a large swelling.

In comparing **Man** with other **mammals**, it is easy to see characters in which he differs strikingly from the lower forms, but when one tries to separate him from the higher forms, such as the Anthropoid apes (Orangs, Chimpanzees, and Gorillas), one finds that the points of difference are comparatively few. It is true that he alone of the Anthropoidea is unable to oppose his big toe to the other toes of that foot, and that whilst Apes sometimes run on all fours he invariably walks erect; he plants the soles of his feet flat on the ground; he also has a better developed heel than any Monkey has, and his arms are shorter than his legs, the latter growing faster after birth than the former. None of these differences can however be looked upon as of much importance.

The only essential difference is, that in Man the brain is much more highly developed than it is in the highest Apes. Thus, whilst the average weight of the human brain is forty-eight ounces, the Gorilla never possesses a brain exceeding twenty ounces in weight. With this greater development of brain goes the greater intellectual capacity of the human being.

It is certain that man cannot lay claim to having arisen from an ancestral Ape similar to any of the known forms; the most that can be said is that he and they have probably a common ancestor, but that in the struggle for existence he has out-stripped them and has attained a much higher condition of intellectual development.

INDEX AND GLOSSARY.

A

Acarus scabiei, 114

Acetabulum, 125

Achromatin (α, not; χρῶμα, colour), 20

Aërobic (ἀήρ, air; βίος), applied to Bacteria, for the growth of which free oxygen is necessary, 171, 175

Acorn-shells, 104

Actinozoa, 87; see *Sea-anemones.*

Albumen gland, in Tapeworms, 226

Aleurone grains, 18

Algæ (*alga*, seaweed), Thallophytes containing chlorophyll, 63

Alimentary or Digestive System—
  in Craniata, 131-134
  in Cœlenterata, 83
  in Crustaceans, 102
  in Earth-worms, 94
  in Echinodermata, 97, 99, 100
  in Liver-fluke, 198
  in Gasteropoda, 118
  in Hydra, 194
  in Hydrozoa, 87, 89
  in Insecta, 108
  in Lamellibranchiata, 116
  in Leech, 242
  in Sea-anemones, 89
  in Sponges, 85

Alternation of Generations—
  in Bryophytes, 67
  in Hydrozoa, 88
  in Phanerogamia, 76
  in Pteridophytes, 70, 71
  in Tape-worms, 214

Amphioxus (ἀμφί, on both sides; ὀξύς, sharp), 121
  development of, 30, 32
  formation of mesoblast in, 37

Amœba (ἀμοιβός, to change), 7, 84, 156-160
  anatomy of, 157
  digestion in, 159
  figure of, 157
  physiology of, 158
  reproduction of, 160

Anabolism (ἀναβολή, that which is thrown up), the constructive processes, by which the protoplasm is built up, 5

Anal respiration—
  in Crustacea, 104
  in Insecta, 108

Anaërobic (α, not; ἀήρ, air; βίος, life), applied to Bacteria for the growth of which free oxygen is not necessary, 171, 175.

Anatomy (ἀνατομή, dissection)—
  of Amœba, 157, 158

Anatomy (*continued*)—
of Dog-fish, 253
of Gregarina, 184, 185
of Hydra, 190
of Leech, 242
of Liver-fluke, 198
of Protococcus pluvialis, 161
of Protococcus viridis, 164
of Round-worms, 229
of Tape-worms, 218
of Vorticella, 177
of Yeast plant, 161
Ancylostomum duodenalis, 234, 235
figure, 235
Andrœcium (ἀνηρ, a male; οἶκος, a house), the male organs of a flower, 74
Angiosperms (ἀγγεῖον, a vessel; σπερμα, a seed), fertilisation and development of, 78
Anguillulidæ, 230
Annelids (*annellus*, a little ring), 91, 94
Anther, 74
Antheridium, male organ in the lower plants, 71
Antherozoid, the male reproductive cell in the lower plants.
Anus, the posterior opening of the alimentary canal.
Apex (growing point)—
of Angiosperms, 60
of Ferns, 61, 73
of Mosses, 61
Apus, 105
Arachnoidea, 102, 105, 112-116
circulatory system of, 113
figures, 113, 114, 115
respiratory system of, 112, 114
spinning glands of, 113
stinging organ of, 112

Archegonium (ἀρχέγονος, the first of a race), female organ in lower plants, 1
Archenteron (ἀρχε, chief; ἐντερον, an intestine), cavity of diblastula, 31
Area opaca, 35
Area pellucida, 35
Armadillo, 260
Arrow-worms, 96
Arthropoda (ἄρθρον, a joint; πούς, a foot), 102
Arthrospore (ἄρθρον, a joint; σπορά, a seed), in Bacteria, 173
Ascaris lumbricoides, 231-233
figure, 231
Asci (ἀσκός, a bag), spore-bearing branches of certain fungi, 65
Ascospores, spores borne on asci, 65
Auditory sense capsules, 130
Auricle (*auricula*, the outer ear), 140, 142
Axial cylinder of Bryophytes, 69
Axis cylinder, 50

B

Bacilli (*bacillum*, a little stick), 173, 174
Bacteria (βακτήριον, a little staff), 170-175
anatomy of, 170
diseases due to, 172
fermentation due to, 171
figure of, 173
putrefaction due to, 171
reproduction of, 172
zooglœa condition of, 172
Balanoglossus, 121
Barnacles, 102, 104, 105
Bast, 56
fibres, 56

Bifid (*bis*, twice ; *findo*, I cleave), forked, divided into two.

Bilharzia hæmatobia, 91

Biology (βίος, life ; λόγος, a discourse), the science of life, 1

Bioplasm (βίος, life ; πλάσμα, anything moulded), see *Protoplasm.*

Birds, appendages of, 127

Blastoderm, blastodermic area (βλαστός, a bud ; δερμα, skin), 31, 32
figure, 34

Blastopore (βλαστός, a bud ; πόρος, a passage), the opening of the diblastula to the exterior, 31

Blastula (dim. of βλαστός, a bud), 29
figures, 31, 32

Blood, 51, 138

Body-cavity, formation of, 37, 38
in Cœlomata, 83
in Earth-worm, 94
in Leech, 248
in Liver-fluke, 200
in Nematodes, 230
in Round-worms, 230
in Tape-worms, 222
in Vertebrates, 123, 139

Bone, 43, 45
figure, 46

Botany (βοτάνη, a herb), the part of Biology which relates to plants.

Bothriocephalus latus, 215

Branchiæ (βράγχια, gills), 135

Brain, 146

Bristle-bearing worms, 94

Bryophytes (βρύον, moss ; φυτόν, a plant), 63, 66-70; alternation of generations in, 67 ; axial cylinder of, 69

Bundle sheath, 57, 60

C

Cæcum (*cæcus*, blind), 135

Calyx (κάλυξ, the cup of a flower), the whorl of sepals in Angiosperms, 75

Cambium (*cambio*, to change), 56
fascicular, 56
interfascicular, 56

Capillaries (*capillaris*, like a hair), 143

Carbo-hydrates, 5

Carnivorous plants, 81

Carpel (καρπός, fruit), 75

Carpus (καρπός, the wrist), 127

Cartilage, 43, 45
figure, 46

Cell (*cella*, a chamber), 6
figures, 7
history of, 19

Cell-division, 20, 23
direct and indirect, 20
figure, 21

Cell-membrane, 10

Cell-sap, 6, 13
description of, 15

Cell-wall, 6, 13, 15
animal cell, 13
figure, 10
plant cell, 13-19
figures, 11, 14
typical cell, 10-13

Cellulose, substance of which plant cell-walls are composed, 13
absence in plants, 81
presence in animals, 81

Centipedes (*centum*, a hundred ; *pes*, a foot), 106

Central corpuscles, 22

Cephalochorda (κεφαλή, the head ; χορδή, a chord), 121

Cephalopoda (κεφαλή, the head ; πούς, a foot), 116, 119

Cephalopoda (*continued*)—
　nervous system of, 119
　respiratory system of, 119
　sensory organs of, 119
　vascular system of 119
Cephalo-thorax (κεφαλή, the head ;
　θώραξ, the chest)—
　in Arachnoidea, 112
　in Crustacea, 103
Cercariæ (κέρκος, the tail)—
　of Distomum hæmatobium, 91
　of Liver-fluke, 206
Cerebral hemispheres, 147
Cestodes (κεστός, a studded girdle),
　91, 92, 208
Cetacea (*cetus*, a whale), 259
Chætopoda, 94
Chara, 66
Chick, development of, 31
Chitin, horny substance present in
　many epidermal structures.
Chlorophyll(χλωρός, green ; φύλλον,
　a leaf), green colouring matter
　of plants, 15
　figure, 18
　absence in plants, 80
　presence in animals, 81
Chrysalis (χρυσός, gold), 111
Cilia (*cilium*, an eyelash), a deli-
　cate motile process of proto-
　plasm ; a flagellum.
Circulatory system—
　in Craniata, 138-144
　in Crustacea, 103
　in Echinodermata, 97, 100
　in Insecta, 108, 109
　in Lamellibranchiata, 118
　in Leech, 247
　in Earth-worms, 94
Cirrus-sac (*cirrus*, a lock of curled
　hair), in liver-fluke, 201, 225
Clitellum (*clitella*, a pack-saddle),
　in Leech, 242

Cloaca (*cloaca*, sewer), the common
　sewer, a cavity into which the
　alimentary and genito-urinary
　canals open, 135
Club-moss, see *Pteridophytes.*
Cnidoblast (κνίδη, a nettle ; βλαστός,
　a bud), in Hydra, 192
Cnidocil (κνίδη ; *cilium*, an eyelash),
　in Hydra, 192
Coccidia, 183
Cocoon (κόγκη, a shell), 111
Cœlenterata, 83, 87-90
　alternation of generations in,
　88
Cœlenteron (κοίλος, hollow ; ἐντε-
　ρον, an intestine), see *Archen-
　teron*, 38
Cœlomata, 84, 90-119
Cœlome (κοίλωμα, a hollow), see
　Body-cavity.
Collenchymatous tissue,see *Tissues.*
Conceptacle (*conceptaculum*, a place
　of conception), 66
Conidia (κόνις, dust), spores found
　in some fungi, 65
Conjugation (*cum*, together; *jugum*,
　a yoke), the union of two cells
　for the purpose of reproduction.
　in Amœba, 160
　in Gregarina, 186
　in Vorticella, 181
Connective tissue, see *Tissues.*
Contractile or pulsating vacuole,
　158, 180
Convoluta Roscoffensis, 81
Coracoid, 125
Coral, 89, 90
Corium or true skin, 130
Cork cambium, 58
Corpora quadrigemina, 148
Corpuscle (*corpusculum*, dim. of
　corpus, body), 51, 138
Cortex (*cortex*, bark), 55

Corolla (*corolla*, a little wreath), the inner whorl of the perianth of Angiosperms, made up of petals, 75

Cotyledons (κοτύληδιον, a cup), the seed leaf of the vegetable embryo, 78

Crabs, 102, 105

Cranial nerves, 150

Craniata (*cranium*, a skull), 121-155
  alimentary system of, 131-134
  appendages of, anterior, 125-127; posterior, 127, 128
  circulatory system of, 138-144
  excretory system of, 134, 135, 144-146
  figures of, 124, 126, 128, 129, 132, 134, 136, 138, 139, 141, 143, 145, 146, 147, 148, 149, 151, 153, 154
  lymphatic system of, 144
  nervous system of, 146-150
  reproductive system of, 152-155
  respiratory system of, 135-138
  sense organs of, 150-152
  skeleton of, 121-130
  visceral skeleton of, 123

Cranium (κράνιον, the skull), 129

Crayfish, 104, 105

Crop, 133

Crustacea (*crusta*, the rind, crust), 102, 105
  alimentary system, 102
  appendages of, 102, 103
  blood vascular system of, 103
  cephalo-thorax of, 103
  exoskeleton of, 102
  figures of, 103, 104
  reproduction and development of (metamorphosis), 105
  respiratory system of, 104

Crypotogamia (κρυπτός, hidden; γαμός, nuptials), flowerless plants, 70

Crystalloids, 18

Cuticle (dim. of *cutis*, skin), 54, 179, 184, 192, 221, 229

Cuttle-fishes, 116

Cysticercus (κύστις, a bladder; κερκος, a tail), in tape-worms.

D

Demodex folliculorum, 115
  figures, 114

Development, 23 *et seq.*
  of Crustacea, 105
  of Gregarina, 186, 187
  of Hydra, 195, 196
  of Insecta, 110, 111
  of Liver-fluke, 203-207
  of Phanerogamia, 76-78
  of Skull, 128
  of Starfish, 100
  of Tape-worms, 210-218

Diaphragm (διαφραγμα, a partition wall), 137

Diblastula condition of embryo (dim. of βλαστός, a bud), 30. See *Gastrula*.

Dicotyledons (δίς, twice; κοτύληδιον, a cup), 78

Differences between plants and animals, 79, 82

Digestive system, see *Alimentary system.*

Dipnoi (δίς, twice; πνοή, breath), 135

Disc, in Vorticella, 179

Distomum hæmatobium, 91
  figure, 92

Dog-fish, 122
  alimentary system of, 252
  anatomy of, 252
  circulatory system of, 253

Dog-fish (*continued*)—
 excretory system of, 253
 nervous system of, 254
 reproductive system of, 253
 respiratory system of, 253
 sense organs of, 255
Dragon-flies, 110
Drosera, 81, 82
Duckmole, 256

E

Ear, 152
Earth-worms, 94-96
 figures, 93, 95
Echidna, 256
Echinodermata (ἐχῖνος, a hedgehog;
 δερμα, the skin), 96-100
 alimentary system of, 99
 blood vascular system of, 100
 calcareous skeleton of, 99
 excretory system of, 100
 figures of, 97, 98, 101
 nervous system of, 99
 reproductive system of, 100
 respiratory system of, 100
Echiuridæ, 94
Ectoplasm (ἐκτός, outside), peripheral layer of protoplasm, 13
 in Amœba, 157
 in Gregarina, 184
 in Vorticella, 176
Ectoderm (ἐκτός, outside; δέρμα, the
 skin), see *Epiblast.*
Ectosarc (ἐκτός, outside; σαρξ,
 flesh), see *Ectoplasm.*
Egg-cell, see *Ovum.*
Embryo (ἐμβρυον, a fœtus), the
 name given to a young animal
 or plant whilst it is still unable to lead an independent
 existence.
Embryo sac, 75

Encystment—
 in Gregarina, 186
 in Protococcus, 165
 in Vorticella, 182
 in Tape-worms, 211
Endoderm (ἔνδον, within; δερμα,
 skin), 30. See *Hypoblast.*
Endogenous spore formation,(ἔνδον;
 γενναώ, to produce), 161, 162
Endonucleus (ἔνδον, within), 12
Endoplasm (ἔνδον, within), internal
 portion of protoplasm, 13, 157
 in Amœba, 157
 in Gregarina, 184, 185
 in Vorticella, 176
Endosarc (ἔνδον, within; σαρξ,
 flesh), see *Endoplasm.*
Endosmosis (ἔνδον, within; ώσμός,
 a thrusting in), diffusion of a
 liquid through a membrane,
 from without inwards, 183
Enteron (ἔντερον, an intestine), 190
Entomostraca, 105
Epiblast (ἐπί, upon; βλαστὸς, a
 bud), the outer of the three
 primitive germ layers, 30
 structures derived from, 38
 figure, 39
Epidermis (ἐπί, upon; δερμα, the
 skin), 53, 130
Epididymis (ἐπί, upon; δίδυμοι, the
 testicles), 155, 254
Epiglottis, 137
Epithelial tissue, see *Tissues.*
Ethmoid bones, 130
Eurotium—
 figure, 64
 reproduction of, 64, 65
Eustachian tubes, 133
Excretory system—
 in Amœba, 159
 in Craniata, 134, 135, 144-146
 in Earth-worms, 94

Excretory system (*continued*)—
  in Echinodermata, 100
  in Insecta, 110
  in Lamellibranchiata, 118
  in Leech, 245
  in Liver-fluke, 200
  in Vorticella, 180
Exoskeleton—
  in Crustacea, 102
  in Armadillo, 260
Eye, 152

F

Fæces (*fæx*, sediment), 134, 135
Falciform body (*falx*, a sickle; *forma*, resemblance), in Gregarina, 187
Fats, 5
Fat tissue, 46
Femur (*femur*, the thigh), 127
Fermentation due to Bacteria, 171
Fern, see *Pteridophytes.*
Fertilisation, 23-28
  figure, 27
Fertilising canal, in Tape-worms, 226
Fibrovascular bundles (*fibra*, a small thread; *vasculum*, a little vessel), 53, 55
  figure, 57
Fibula, 127
Filament, 74
Filaria medinensis, 238
  figure, 239
Filaria sanguinis hominis, 238-240
Fins, 124
Fish, appendages of, 127
Fission—
  in Amœba, 160
  in Bacteria, 172
  in Gleocapsa, 168
  in Hydra, 194
  in Protococcus, 164
  in Vorticella, 180, 181

Fission fungi, see *Bacteria*
Flagella (*flagellum*, a whip)—
  in Bacteria, 172
  in Hydra, 193
  in Protococcus, 165
Flat-worms, 91
Flowering plants, see *Phanerogamia*
Food vacuoles—
  in Amœba, 158
  in Vorticella, 180
Frog, 133
Frontal bone (*frons*, the forehead), 130
Fucus, Bladder wrack, 66
  figure, 65
Fungi (*fungus*, a mushroom), Thallophytes containing no chlorophyll, 63

G

Gall-mites, 115
Ganglion (γαγγλιον, a swelling), an enlargement in the course of a nerve.
Ganglion cells, 49
Gasteropoda (γαστήρ, the stomach; πους, a foot), 116, 118
  alimentary system, 118
  respiratory system, 118
Gastrula condition of embryo (γαστήρ, the stomach), condition in which the embryo consists of a two-layered sac, 30
Gemmation (*gemma*, a bud)—
  in Hydra, 194
  in Hydrozoa, 87, 88
  in Sponges, 86
  in Yeast plant, 161, 162
Genital cloaca, in Tape-worms, 227
Genital pore, in Tape-worms, 227
Germ-cell, see *Ovum Spermatozoon*, and *Sperm Mother-cell.*

Germ-disc, 29
  figures, 29, 30, 33
Germinal spot (nucleolus of ovum), 24
Germinal vesicle (nucleus of ovum), 24
Gills, 123
Gill-slits, 123
Gland, a collection of cells forming a secretory or excretory organ.
Glenoid cavity, 125
Gleocapsa, 168, 169
  anatomy of, 168
  physiology of, 169
  reproduction of, 168
Glomerulus (dim. of *glomus*, a ball of thread), 145
Glottis, 137
Gordiidae, 230
Gregarinæ (*gregarius*, belonging to a flock), 183-187
  anatomy of, 184, 185
  figure of, 185
  reproduction of, 186, 187
Gregarinidæ (*gregarius*, belonging to a flock), 84, 183
Ground tissue, see *Tissues*.
Growing point, see *Apex*.
Guard cells, 54
Gymnosperms (γυμνός, naked; σπέρμα, a seed), 78
  fertilisation and development of, 76-78
  figure of, 77
Gynœcium (γυνή, a female; οἶκος, a house), the female organ of a flower, 75

H

Hæmoglobin (αἷμα, blood; *globus*, a round body), 138
Hæmatococcus pluvialis, see *Protococcus pluvialis*, 164

Hairs—
  animal, 38, 131
  plant, 53
    figure of, 53
Harvest mites, 115
Heart—
  in Craniata, 139
  in Crustacea, 104
  in Dog-fish, 253
  in Lamellibranchiata, 118
Hepatic portal circulation, 141
Hermaphrodite (Ἑρμῆς, Mercury; Ἀφροδίτη, Venus), applied to animals and plants in which both male and female organs are present.
Hemichorda, 121
Hermit-crab, 105
Hirudinæ, 94, 96
Hirudo medicinalis, see *Leech*.
Horsetails, see *Pteridophytes*.
Host, the organism in which a parasite lives.
Humerus, 127
Hyaloplasm (ἴαλος, glass), clear part of protoplasm, 12, 13
Hydatid (ὕδατις, a watery vesicle), of Tape-worms, 211
Hydra (ὕδρα, a water serpent), 188-196
  anatomy of, 190
  digestive system of, 194
  figures of, 189, 191
  fusca, 190
  muscular system, 192,
  presence of chlorophyll in, 81
  reproductive system in, 194
  viridis, 81, 190
  vulgaris, 190
Hydrozoa, 87-89
  alimentary system of, 87
  alternation of generations of, 88
  figure, 88

Hydrozoa (*continued*)—
nervous system of, 89
reproductive system of, 87, 88
Hyoid arch, 130
Hyoid bone, 130
Hyphæ (ὑφαίνω, to weave), separate filaments of a fungus
aerial, 65
mycelial, 65
submerged, 64
Hypoblast (ὑπό, under; βλαστός, a bud), the inner layer of the three primitive germ layers, 30
structures derived from, 39
Hypostome (ὑπό, under; στόμα, the mouth), in Hydra, 191

I

Ilium, 125
Imago (*imago*, a likeness), 111
Infusoria, 84. See *Vorticella*.
Insecta, 102, 105, 106-112
alimentary system of, 108
appendages of, 108
circulatory system of, 108, 109
excretory system of, 110
figures, 106, 107, 109, 111
metamorphosis of, 110, 111, 112
muscular system of, 108
nervous system of, 108
reproductive system of, 110
respiratory system of, 107, 108
sensory organs of, 107, 108
Integument—
of Craniata, 130
of Tape-worms, 221
Intercellular digestion of Hydra, 194
Intercellular substance, 44
Intestine, 133, 134

Intracellular digestion—
in Amœba, 159
in Hydra, 194
in Vorticella, 180
Invagination (*in*, into; *vagina*, a sheath), the pushing in of one part into another, 30
Invertebrate, 83-119
Ischium, 125

J

Jaws, 130
Jelly-fish, 89

K

Kangaroo, 256
Karyokinesis (κάρυον, a kernel; κίνησις, a movement), changes which take place in the nucleus, 20
Katabolism (καταβολή, a laying down), the destructive processes, by which the protoplasm is broken down into simpler substances, 5
Keimplasma (*keim*, germ), 24
Kidneys, in Craniata, 144, 145
King crab, 112, 115

L

Lachrymals, 130
Lacteals (*lac*, milk), 144
Lamellibranchiata (*lamella*, a plate; βράγχια, gills), 116-118
alimentary system of, 116
circulatory system of, 118
excretory system of, 118
figure of, 117
muscular system of, 116
nervous system of, 118

Lamellibranchiata (*continued*)—
   reproductive system of, 118
   respiratory system of, 117
   water vascular system of, 117
Larva (*larva*, a mask), the young
   of Insecta, etc., in their first
   stage of metamorphosis.
Larynx, 137
Leaves, structure of, 62
Leech, 94
   alimentary system of, 242
   anatomy of, 242
   excretory system of, 245
   nervous system of, 250
   reproductive system of, 249
   vascular system of, 247
Lichen, 66
Limulus, 115
Liver, 133, 134
Liver-fluke—
   alimentary system of, 198
   anatomy of, 198
   excretory system of, 200
   figures of, 199, 202, 204
   life history of, 203
   nervous system of, 200
   reproductive system of, 201
Liverworts, 67-70
Lobsters, 102, 105
Lob-worms, 96
Lungs, 137
Lymph (*lympha*, water), 51, 144
Lymphatic system in Craniata, 144

M

Macrosporangium, 75
Macrospore (μακρος, great ; σπορά,
   a seed), a spore, from which
   a female prothallium develops,
   71, 166
Madreporic plate, 98
Malocostraca, 105
Malpighian body, 146

Mammalia (*mamma*, a breast)
   alimentary system of, 261
   dentition of, 260
   excretory system of, 264
   reproduction of, 264
   respiration of, 262
Mandibular arch, 130
Manus, 127
Marsupialia (*marsupium*, a pouch
   256
Maxillæ, 130
Mayflies, 110
Medulla oblongata, 148
Medullary rays, 56
Medullary sheath (*medulla*, mar-
   row), 50
Medusoid (Μέδουσα, one of the
   Gorgons), 87
Mehlis' body, 226
Meristem tissue (μεριστός, divided),
   see *Tissues.*
Mesenteron (Μέσος, middle ; ἔν-
   τερον, an intestine), 38
Mesoblast (Μέσος, middle ; βλαστὸς,
   a bud), the middle of the three
   primitive germ layers, 37
   figure, 38
   structures derived from, 39
Meso-nephros, 155
Mesophyll, spongy, 62
Meso-thorax, 108
Metabolism (μεταβολή, a change),
   the chemical changes which
   take place during the life history of protoplasm. See
   Anabolism and *Katabolism,* 5
Metamorphosis—
   of Crustacea, 105
   of Insecta, 110, 111, 112
Metathorax, 108
Metazoa (μετά, after ; ζῶον, an
   animal), animals whose bodies
   consist of more than one cell, 83

Mesoglœa (μέσοι, middle ; γλοία, glue), 192

Micrococci, 173

Microsporangium, 74, 75

Microspore (μικρος, small ; σπορά, a seed), a spore from which a male prothallium develops, 71, 166

Millipedes, 106

Mites, 112, 114, 115

Mollusca, 116-118

Monocotyledons (μόνος, single ; κοτύληδιον, a cup), 78

Monotremata (μόνος, single ; τρῆμα, a hole), animals which have only one external opening, 256

Morula (dim. of *morum*, a mulberry)—
hollow, 29
solid, 29

Mosses, 66-70
figure, 68. See *Bryophytes*.

Motor root, 150

Moulds, 64

Müllerian duct, 154

Muscle—
smooth or non-striated, 47
striated, 47

Muscular system—
in Echinodermata, 99
in Hydra, 192
in Insecta, 108
in Lamellibranchiata, 116

Muscular tissue, see *Tissues*.

Mussel, 116-118

Mycellium (μύκης, a fungus), a felt-like mass formed of interwoven hyphæ, 65

Myophan striation (μύς, muscle ; φαίνω, to appear), 180

Myriapoda, 102, 105, 106

Mysis stage, 105

N

Nasal sense capsules, 130

Nasal sense organs, 152

Nauplius, 105

Nemathelminthes, 91

Nematocyst (νῆμα, a thread ; κυστις, a bag), in Hydra, 192

Nematodes (νῆμα, a thread), 94, 229-240
alimentary system of, 229
development of, 230, 231
excretory system of, 229
muscular system of, 229
nervous system of, 229
reproductive system of, 229
sensory system of, 230

Nephridium (νεφρός, a kidney), in worms, 90, 93, 94

Nemertines, 92, 93

Nerve cells, 49

Nervous system—
in Craniata, 146-150
in Echinodermata, 97
in Hydrozoa, 89
in Insecta, 108
in Liver-fluke, 200
in Worms, 90, 93, 94

Nervous tissue, see *Tissues*.

Nitella (*nitio*, to shine), 66
figure, 67

Notochord, (νῶτος, the back ; χορδή, a string), a solid rod of cells, pinched off from the dorsal region of the hypoblast, during development in vertebrates, 122
sheath of, 120, 122

Nucellus (dim. of nucleus), 78

Nucleolus, 6

Nucleus (*nucleus*, a kernel), 6, 11, 12
aster stage of, 22
coil stage of, 20

Nucleus (*continued*) —
  diaster stage of, 22
  figure, 13
  resting, 20

O

Occipital bone, 130
Œsophagus, 133
Olfactory lobes, 147
Oligochæta, 94
Opossum, 256
Optic lobes, 148
Ornithorynchus, 256
Ovary (*ovum*, an egg) —
  in Craniata, 153
  in Dogfish, 253
  in Earth-worm, 95
  in Echinodermata, 100
  in Hydra, 194, 195
  in Insecta, 110
  in Leech, 250
  in Liver-fluke, 202
  in Mollusca, 118
  in Phanerogamia, 75
  in Tape-worm, 226
Oviduct (*ovum*, an egg; *duco*, I
  lead), the tube which conveys
  ova from the ovary to the
  exterior.
Ovule (dim. of *ovum*), 75
Ovum (*ovum*, an egg) —
  description of, 24
    figures, 10, 27, 32
  segmentation of, 28
    figures, 28-36
Oxyuris vermicularis, 233
  figure, 232

P

Pancreas (πᾶν, all; κρεας, flesh),
  134
Parenchymatous tissue, see *Tissues*.

Parachordals (παρά, near ; χορδή. a
  string), 130
Paranucleus (παρά, beside; nucleus),
  180
Parasite (παράσιτος, one who lives
  at other people's tables).
Parietal bones (*paries*, a wall), 130
Parthenogenesis (παρθένος, a virgin;
  γένεσις, origin), 23
Pectoral girdle, 124, 125
Pedicelli, 97
Pelvic girdle, 125
Penæus, 105
Penicillium (*pēnicillium*, a brush),
  reproduction of, 64
Pentastomata (πέντε, five ; στόμα,
  a mouth), 115
  figure, 115
Pericardium (περί, about ; καρδία,
  the heart), 138
Peripatus, 106
Peristome (περί, around ; στόμα,
  the mouth), 178
Peritoneal cavity, the part of the
  body-cavity below the dia-
  phragm.
Permanent tissue, see *Tissues*.
Pes (*pes*, foot), 127
Petals (πέταλον, a leaf), 74
Phanerogamia (φανερόω, to make
  manifest ; γαμός, nuptials),
  63, 73-78
  fertilisation and development of,
  76-78
  figures, 73, 74, 75, 76, 77
Pharynx (φαρυγξ, the throat), 131
Phloëm (φλοίος, bark), 56
Phycocyan, 16
Phycoerythrin, 16
Phycophæin, 16
Physiology (φύσις, nature ; λόγος, a
  discourse) —
  of Amœba, 158, 160

Physiology (*continued*)—
    of Hydra, 193
    of Protococcus pluvialis, 166
    of Protoplasm, 4,
    of Yeast plant, 162
Pineal gland, 147
Pistil, 75
Pith, 56
Pituitary body (*pituita*, mucus), 147
Placentalia (πλακοῦς, a flat cake), 258
Plant structure, 62-78
Plathelminthes, 91
Plexus (*plecto*, to twist, or weave).
Plumule (*plumula*, a little feather), 78
Polar bodies, or vesicles, 25
    figure, 25
Polian vessels, 99
Pollen grain, 75
Pollen sac, 75
Pollen tube, 76, 77
Polychæta, 96
Polycystidia, 184
Polypite, 87, 89
Pons varolii, 148
Porifera, 85
Porpoise, 259
Præcoracoids, 125
Præmaxillæ, 130
Primary vesicles, 147
    first, 147
    second, 147, 148
    third, 148
Primitive streak, 34
    figure, 34
Proctodeum (πρωκτός, the anus; ὁδαιός, belonging to a way), 131
Pro-echidna, 256
Proglottis, 208
Pronephros, 154

Pronucleus—
    female, 25
    male, 27
Proscolex, bladder-worm, 211
Prosenchymatous tissue, see *Tissues*.
Proteids (πρῶτος, first), 4
Prothallium (πρό, before ; θαλλός, a twig), the sexual generation of Pteridophytes, etc., 71
Pro-thorax, 108
Protococcus pluvialis, 164-169
    anatomy of, 165
    physiology of, 166
    reproduction of, 165
    respiration of, 167
Protococcus viridis, 164
    anatomy of, 164
    figure, 165
    reproduction of, 164
Protoplasm (πρῶτος, first ; πλάσμα, anything moulded) —
    characters of, 3
    chemical composition of, 4
    effects of changes of temperature, electricity, etc., on, 8
    history of, 8
    movements in, 6
    need of water, 8
    presence in plant cell, 13
    respiration of, 7
Protozoa (πρῶτος, first ; ζῷον, an animal), animals whose bodies consist of one cell only, 83
Protracheata, 102, 105, 106
Pseudonavicella (ψευδής, false ; navicula, a little boat), 186
Pseudopodium (ψευδής, false ; πους, foot), 157, 158
Psorosperms, 84, 183
Pteridophytes, 63, 70-73
    alternation of generations in, 70, 71

Pteridophytes (*continued*)—
  figures, 69, 70, 71, 72
  fibrovascular bundle of, 72
  leaf of, 70, 73
  root of, 73
  stem of, 73
Pubis, 125
Pulmonary circulation, 142
Pulsating vacuole, 158, 180
Pupa, 111
Putrefaction due to Bacteria, 171
Pyrenoid, 165

R

Rabbit, reproduction of, 35
Radicle (*radic*, a root), 78
Radius, 127
Receptaculum seminis, 95, 110
Rediæ—
  of Distomum hæmatobium, 91
  of Liver-fluke, 206
Renal portal circulation, 141
Reproduction, 23-39
Reproductive system—
  in Amœba, 160
  in Bryophytes, 67
  in Cœlenterata, 87-90
  in Craniata, 152-155
  in Crustacea, 105
  in Earth-worm, 91, 92, 94, 95
  in Echinodermata, 100
  in Gregarina, 186, 187
  in Hydra, 194
  in Hydrozoa, 87-89
  in Insecta, 110
  in Lamellibranchiata, 118
  in Liver-fluke, 201-203
  in Phanerogamia, 74-78
  in Protococcus pluvialis, 166
  in Protococcus viridis, 164
  in Pteridophytes, 71
  in Sea-anemones, 90
  in Sponges, 85, 86, 87

Reproductive system (*continued*)—
  in Tape-worms, 240
  in Thallophytes, 65
  in Vorticella, 180-182
  in Yeast plant, 161-162
Respiratory system—
  in Craniata, 135-138
  in Crustacea, 104
  in Earth-worm, 93, 94
  in Echinodermata, 100
  in Gasteropoda, 118
  in Insecta, 107, 108
  in Lamellibranchiata, 117
  in Protococcus pluvialis, 167
  in Tracheata, 105, 107, 116
Reticulum, 12, 13
Rhizoid (ῥίζα, root; εἶδος, form),
  66
Rhizopoda, 84
Ribs, 128
Ribbon-worms, 92
Ringed-worms, 94
Root—
  cap, 60
    figure, 61
  growing point of, 60
  structure, 59, 60
Rotifers, 96
Round-worms, 91, 94

S

Saccharomyces cerevisiæ (σάκχαρον,
  sugar; μύκης, fungus), 161
Sagitta, 96
Salivary glands, 131
Saprophytes (σαπρός, putrid;
  φυτον, a plant), Thallophytes
  obtaining their nourishment
  from decomposing organic
  matter, 66
Sarcoptes hominis, 114
Scapulæ, 125

Scolex (σκώληξ, a worm), 215
Sea-anemones, 87, 89, 90
Sea-cows, 258
Sea-cucumbers, 96
Sea-spiders, 112
Sea-urchins, 96
Segmentation (*segmentum*, a piece cut off), 23, 28
  figure, 28, 30
  complete, holoblastic, 29
  partial, meroblastic, 29
Segmentation or cleavage cavity, 29
Sense organs, 150
Sensory root, 150
Sepals (*separ*, separate), 74
Sheep-rot, produced by Liver-fluke, 197, 207
Shell gland—
  in Liver-fluke, 203
  in Tape-worms, 226
Shrimps, 102
Sieve tubes, 56
Sinus venosus, 140
Skeleton, 121-130
Skull, 129
Sirenia, 258
Snails, 116
Snakes, 124
Somatopleure (σῶμα, the body; πλευρὸν, the side), 38
Somite (sῶμα, the body), 242
Sori (σωρος, a heap), a collection of sporangia on the under surfaces of fern leaves, 70
Spermatophores (σπέρμα, a seed; φορέω, to carry), 249
Spermatozoon (σπέρμα, a seed; ζῶον, animal), 25, 26
  figure, 26
Sperm mother-cell, 25
Sphenoid bone, 130 .
Spiders, 102, 105, 112, 113, 114

Spinal cord, 145, 146
Spinal nerves, 150
Spiracle, 252
Spirilla, 173
Spirogyra (*spira*, a coil; *gyrus*, a revolution), 63
Splanchnopleure (σπλαγχνον, an intestine; πλευρὸν, the side), 38
Sponge, 85
Spongilla, 86
  reproduction of, 86, 87
Spongioplasm, the reticulum, or denser part of protoplasm, 12
Sphincter (σφίγγω, I bind).
Sporangium (σπορά, a seed; ἀγγεῖον, a vessel), a spore case, 70
Spores (σπορα, a seed), an asexual reproductive cell.
Sporocyst (σπορά, a seed; κυστις, a bladder), 206
Sporozoa, 84
Squamosals, 130
Stamens (*stamen*, a thread), 74
Starch, 18
Star-fishes, 96
Stem—
  tissues of, 5
  figures, 54, 55, 57
  growing point of, 60
Sternum, 128
Stigma (στίγμα, a spot), 75
Stomach, 133
Stomata (στόμα, a mouth), apertures in the epidermis of a leaf, 54
  figure, 54
Stomodœum (στόμα, mouth; οδαιός, belonging to a way), 131
Stone canal, 98
Strobila, 208

Structureless or supporting lamella (*lamella*, a thin plate), 192
Style (*stylus*, a column), 75
Sundew, 81, 82
Swimming bladder, 135
Symbiosis (συμβίωσις, a living with), 17
in Convoluta Roscoffensis, 81 (note)
in Lichen, 66
Systemic circulation, 142

T

Tadpole, 133
Tænia echinococcus, 215
Tænia mediocanellata, or saginata, 215
Tænia solium, 210, 215
Tape-worms (cestodes), 208-228
anatomy of, 218
cysticercus of, 211
development of, 210
excretory system of, 223
figures of, 208-227
head of, 220
history, 228
hydatid of, 211
life history of, 210
neck of, 220
nervous system of, 221
proscolex of, 211
reproductive system of, 224
Tarsus, 127
Tentacles, 190
Testes, 153
in Craniata, 153
in Dog-fish, 254
in Earth-worm, 94
in Echinodermata, 100
in Hydra, 194, 195
in Insecta, 110
in Leech, 249

Testes (*continued*)—
in Liver-fluke, 201
in Mollusca, 118
in Tape-worm, 225
Thalamencephalon, 147
Thallus (θαλλός, a twig), 63
Thallophytes (θαλλός, a twig; φυτόν, a plant), 63-66
figures, 64, 65
Thread cell, 192
Thymus gland, 133
Thyroid body, 133
Tibia, 127
Tissues, 40-62
animal tissues, 43-51
botryoidal tissue, 248
connective, 43, 45
figure, 45
epithelial, 43
figure, 44
fat, 46
muscular, 43, 47
figures, 48, 49
nervous, 43, 48
plant tissues, 52-62
figures, 53, 54, 55, 57, 61
collenchymatous, 55
ground or fundamental, 53, 54
meristem, 52
parenchymatous, 55, 56
permanent, 52
prosenchymatous, 55
Tongue, 131
Torula, 161
Trabeculæ cranii, 130
Tracheata, 105-116
Tradescantia virginica, 6
Transverse processes, 122
Trematodes, 91
Trichina spiralis, 236-306
figure, 237
Trichocephalus dispar, 233, 234
figure, 234

Trichinosis, 238
Truncus arteriosus, 253
Tunicata, 121
Turbellaria, 91

U

Ulna, 127
Ureters, 145
Urethra, 155
Urinary organs, see *Excretory System.*
Urochorda, 121
Uterus—
    in Craniata, 155
    in Liver-fluke, 202
    in Mammalia, 264
    in Nematodes, 229
    in Tape-worms, 227

V

Vacuoles (*vacuum*, empty), description of, 12, 13
    in Amœba, 158
    in Hydra, 180
    in Plant-cells, 14
    in Vorticella, 177
    in Yeast-plant, 161
Vagina (*vagina*, a sheath), 155
Vascular cryptogams, 70, 73
Vas deferens (*vas*, vessel; *defero*, to carry away), the vessel carrying male generative products to the exterior.
Vasa efferentia (*vas*, vessel; *effero*, to carry off), the efferent ducts of the testis.
Vaucheria, 63
Ventricles of brain, 140
    third, 147
    fourth, 147
    lateral, 147

Ventricles of heart, 140, 142
Vermes, see *Worms.*
Vertebrata, 120, 155
Vertebræ, 122
Vesicula seminalis (dim. of *vesica*, a bladder; *semen*, a seed), 254
Vestibule, 179
Visceral arches, 123
Visceral skeleton, 123
Vitellarian ducts (*vitellus*, yolk), 203
Viviparous (*vivas*, living; *pario*, to beget), born alive.
Vocal cords, 137
Vorticella (dim. of *vortex*, an eddy), 176-182
    anatomy of, 177
    excretory organ in, 180
    figures, 178, 181
    reproduction of, 180-182

W

Water vascular system—
    in Echinodermata, 97, 98, 99, 100
    in Tape-worm, 223
Water fleas, 102, 105
Whales, 259
Wheel animalcules, 96
Wings, 124
Wolffian body, 155
Wood, see *Xylem.*
    fibre cells, 56
    vessels, 56
Woodlice, 102
Worms, 90-96

X

Xylem (ξύλον, wood), the internal part of the fibrovascular bundle, 56

Y

Yeast plant, 161-163
  anatomy of, 161
  physiology of, 162
  reproduction of, 161
Yolk, 27
Yolk gland—
  in Liver fluke, 203
  in Tape-worms, 226

Z

Zooglœa condition of Bacteria (ζῶον, an animal; γλοία, glue), 172
Zoology (ζῶον, an animal; λογος, a discourse), the part of Biology which relates to animals.
Zoospores (ζῶον, an animal; σπορά, a seed), 166

Printed by Hazell, Watson, & Viney, Ld., London and Aylesbury.

www.ingramcontent.com/pod-product-compliance
Lightning Source LLC
Chambersburg PA
CBHW021511210326
41599CB00012B/1211